International Max Planck Research School (IMPRS)
for Maritime Affairs
at the University of Hamburg

For further volumes:
http://www.springer.com/series/6888

Hamburg Studies on Maritime Affairs
Volume 18

Edited by

Jürgen Basedow
Peter Ehlers
Hartmut Graßl
Lars Kaleschke
Hans-Joachim Koch
Doris König
Rainer Lagoni
Gerhard Lammel
Ulrich Magnus
Peter Mankowski
Marian Paschke
Thomas Pohlmann
Uwe Schneider
Jürgen Sündermann
Rüdiger Wolfrum
Wilfried Zahel

Irene Stemmler

The Role of the Ocean in Global Cycling of Persistent Organic Contaminants

Refinement and Application
of a Global Multicompartment
Chemistry-Transport Model

 Springer

Dr. Irene Stemmler
Max Planck Institute for Chemistry
Joh.-Joachim-Becher-Weg 27
55128 Mainz
Germany
irene.stemmler@zmaw.de

Dissertation zur Erlangung der Doktorwürde an der Fakultät für Mathematik,
Informatik und Naturwissenschaften der Universität Hamburg
Erstgutachter: Dr. Thomas Pohlmann
Zweitgutachter: Prof. Dr. Gerhard Lammel
Tag der mündlichen Prüfung: 27. Mai 2009

ISSN 1614-2462   e-ISSN 1867-9587
ISBN 978-3-642-05008-4   e-ISBN 978-3-642-05009-1
DOI 10.1007/978-3-642-05009-1
Springer Heidelberg Dordrecht London New York

Library of Congress Control Number: 2010922989

Cover design: WMXDesign GmbH, Heidelberg

Printed on acid-free paper

Springer is part of Springer Science+Business Media (www.springer.com)

Foreword

The global ocean serves as a repository for long-lived anthropogenic, so-called persistent chemicals. Inputs are riverine runoff and atmospheric deposition. Therefore, pollution is not limited to the marginal seas. Extremely low concentrations are expected in open seawater, but many organic substances tend to accumulate in biota, both fauna and flora, and the more the higher the trophic level along the marine food chain (bio-accumulation). This has caused detrimental effects in marine wild life and poses a hazard for human health. The causative substances are called persistent organic pollutants (POPs). Since recent years the protection of the marine (and total) environment against POPs has been the subject of internationally binding legislation.

Science is developing tools with the aim to predict the behaviour and distribution of old and new (and even non-existing) chemicals and quantify environmental exposure. Persistence implies not only large temporal but also large spatial scales. Many of the substance of highest concern, e.g. a number of obsolete pesticides and flame retardants, have been distributed among environmental compartments worldwide since decades and undergo repeated re-entries into ocean, atmosphere, or soils. The ocean and atmosphere compartments are multi-phase systems of colloidal nature. In conclusion, the perfect model is global but high resolved and capable to describe substance distributions in geographic, compartmental and particle size terms and the mass fluxes between compartments.

The research published in this book uses the presently most comprehensive multi-compartment model, the first which comprises a coupled atmosphere-ocean general circulation model (GCM). GCMs are the state-of-the-art tools used in climate research. The study is on the marine and total environmental distribution and fate of two chemicals, an obsolete pesticide (DDT) and an emerging contaminant (perfluorinated compound) and contains the first description of a whole historic cycle of an anthropogenic substance, i.e. from the introduction into the environment until its fading beyond phase-out.

Hamburg, October 2009 *Gerhard Lammel*

Preface

This book is based on a doctoral thesis submitted to the Faculty of Mathematics, Informatics, and Natural Science at the University of Hamburg in 2009.

The preparation of this thesis would not have been possible without the advice and help I got from many people.

First of all I want to thank Prof. Dr. Gerhard Lammel for guiding me through the entire time with well-balanced motivation and criticism, specific questions, and good-humoured advice, and for giving me the freedom to shape my selfmotivation and independence.

My work and living was financed by the International Max Planck Research School for Maritime Affairs, and during the last month also by the Max Buchner Research Foundation. Thank you very for your trust in me. Thank you also to the directors and coordinators of the IMPRS-MA for providing an intersting education program, which opened my mind for issues I never would got in touch with otherwise.

I am obliged to Dr. Hans Feichter for hosting me in his group, providing me an inspiring working environment and supporting me without any form of reimbursement.

Thanks to all colleagues and friends from the former Aerosol Chemistry group, especially Dr. Francesca Guglielmo for inspiration, education and endless debugging support.

I would like to express my thanks also to Dr. Helmut Haak for technical support, and Dr. Ernst Maier-Reimer, Dr. Jochen Segschneider, and Dr. Katharina Six for valuable comments.

Many thanks to my office mates, Semeena, Álvaro, Angelika, Christine, and Katrin for making office hours the most enjoyable. Special thanks to Chris, whose patient proof reading, editing and drawings helped shaping this manuscript.

Finally, I would like to thank my family and friends, and especially Anni for providing me a loving home free of pollutants.

Hamburg, Germany, 2009   *Irene Stemmler*

Contents

Acronyms

AGG	Experiment identifier (aggregation of marine snow)
AMAP	Arctic Monitoring and Assessment Programme
CDOM	Coloured Dissolved Organic Matter
COG	Centre Of Gravity
COM	Colloidal Organic Matter
CS	Continental Shelf
DOC	Dissolved Organic Carbon
FAO	UN Food and Agriculture Association
GCM	General Circulation Model
HAM	Hamburg Aerosol Model
HAMOCC	Hamburg Ocean Carbon Cycle Model
LD	Lethal Dose
LRT	Long-Range Transport
LRTP	Long-Range Transport Potential
MBM	Multimedia mass balance Box Model
MERIS	Medium Resolution Imaging Spectrometer
MPIOM	Max Planck Institute Ocean and Sea Ice Model
MCTM	Multicompartment Chemistry-Transport Model
MLD	Mixed Layer Depth
NIR	Near InfraRed
NN	Neural Network
OASIS	Ocean-Atmosphere-Sea Ice-Soil coupler
POC	Particulate Organic Carbon
POP	Persistent Organic Pollutant
RGB	Red Green Blue color map
SAT	Experiment identifier (satellite data assimilation)
SOC	Semivolatile Organic Compound
SPM	Suspended Organic Matter
SST	Sea Surface Temperature
UNEP	United Nations Environmental Programme
US EPA	U.S. Environmental Protection Agency

Chapter 1
Introduction

1.1 Persistent organic contaminants

Organic contaminants in the environment constitute a diverse group of organic substances, out of which the persistent organic pollutants (POPs) are of special concern due to their unique properties. Even at low concentrations they are harmful to humans and wildlife. Effects include carcinogenesis, immune dysfunction, neurobiological disorders and reproductive and endocrine disruption [Ritter et al (1995)]. If semivolatile or volatile, persistent pollutants undergo long-range transport, and, hence, are present in remote regions where they have never been produced or used [Patton et al (1989), Iwata et al (1993), [Villa et al (2003)]. Their highly persistent nature results from their ability to resist degradation pathways such as photolysis, hydrolysis, and biodegradation to a high degree [Atkinson et al (1999)]. POPs have a low aqueous solubility, and are liphophilic. Therefore they bioaccumulate in lipid rich tissues and biomagnify through food chains [De Wolf, W. et al (1992), Skei et al (2000)]. The complexity of POPs cycling resulting from these properties is a challenge for fundamental science, but at the same time has significant implications for international chemicals law. Among other legal instruments, the UNEP Stockholm Convention [UNEP (2001)] currently regulates usage and production of the so-called 'dirty dozen' or legacy POPs. A large number of chemicals exist which have properties comparable to the banned POPs. One group of compounds that has generated considerable scientific, regulatory, and public interest on an international scale are the perfluorocarbons (PFCs). The large attention arises due to the fact that they have numerous consumer and industrial applications, and that some of these compounds have been shown to be globally distributed, persistent, and present in humans and wildlife, as well as toxic for laboratory animals, such as rodents [Giesy and Kannan (2002), Olsen et al (2003), Falandysz et al (2007)].

1.2 Global cycling of persistent organic contaminants

Global environmental fate of persistent organic contaminants results from the in-
terplay of various processes including physical transport, partitioning between the
compartments, and biogeochemical cycles. The atmospheric pathway of a contaminant is governed by the spatial distribution of its sources, air circulation patterns,
exchange with the Earth's surface, and chemical transformation. The mass exchange
of contaminants between the atmosphere and the Earth's surface involves dry deposition of particles, gas exchange, precipitation scavenging and sea spray driven generation. Removal by precipitation depends on gas-particle partitioning, solubility of
the vapour phase substances in water and adsorption to snow. Atmospheric transport
can be subdivided into single-hop and multi-hop pathways. A cycle is termed as a
single-hop, if a compound is emitted to the atmosphere, transported and deposited to
the Earths surface, and never returns to the atmosphere. If a compound re-enters the
atmosphere after initial deposition to the Earth's surface, it can move through the
environment by taking multiple atmospheric hops, which is called the 'grasshopper
effect' [Wania and Mackay (1993), Wania and Mackay (1996)]. Processes by which
this can occur include volatilisation from the Earths surface under warmer temperatures when the contaminant was deposited, sudden exposure of ocean water to the
atmosphere after being covered by ice, or resuspension of dust or snow by winds.
Therefore, defining the ultimate distribution of a multi-hop contaminant is not sim-
ply a function of the processes of atmospheric transport, circulation and removal,
but also of surface processes that control its re-entry into the atmosphere after initial
deposition.

Oceans are also important for global cycling of persistent organic pollutants [Gustafs-
son and Gschwend (1997), Dachs et al (2002), Jurado (2006), Guglielmo (2008)].
Their large volumes imply that they may represent an important inventory of POPs.
In the water column, POPs can be found either truly dissolved, sorbed to colloids or
sorbed to particles. They are affected by the hydrodynamics of water masses such
as turbulence and advection of water masses. Dissolved POPs can re-volatilise back
to the atmosphere, whereas particle-bound compounds are subject to gravitational
settling. Thus, the oceans can act as a source, as a storage compartment, and as a
sink. In addition, sediments may exert an influence on the distribution of contaminants in the deeper water column through resuspension and diffusion from sediments porewater [Eggleton and Thomas (2004), Wiberg and Harris (2002)]. The
contribution of oceanic currents to global transport depends on the state of hy-
drophobicity of a compound and its tendency to partition to organic matter. In a
model study northward transport of hexachlorocyclohexane (γ-HCH, also termed as
lindane) and dichlorodiphenyltrichloroethane (DDT) by ocean currents was found
to be small on a global scale [Guglielmo (2008)]. Locally, especially for western
boundary currents such as the Gulf stream, it was shown to be of importance. The
work of Dachs et al (1999) and Dachs et al (2002)) show that phytoplankton uptake
and settling of particulate matter act as drivers of the oceanic sink of polychlorinated biphenyls (PCBs) and dibenzo-p-dioxins and -furans (PCDD/Fs), which are

also classified as POPs. Generally, sinking and export of POPs to the deep ocean has been described as a major removal mechanism from surface waters for the more hydrophobic POPs [Gustafsson et al (1997); Dachs et al (2002)]. Such affirmation, however, rely upon few vertical flux measurements of POPs and on empirical parameterisations. A global model study with a focus on impacts of gravitational settling in the particle phase was conducted by Guglielmo (2008). The model results show that the effect of partitioning to organic matter is spatially very inhomogeneous. Regional migration of persistent compounds into the deep ocean is altered by gravitational settling, but globally ocean dynamics were shown to be of greater importance. It was also suggested that in addition to its importance for the oceanic sink of POPs, the export to deep ocean affects the atmospheric residence time of POPs [Scheringer et al (2004)]. A general assumption is that the organic phase to which pollutants sorb is mainly to phytoplankton, which is supported by studies where PCBs follow the profile of the phytoplankton biomass [Dachs et al (1997)]. Planktonic webs are are at the bottom of the food web, and indeed they may be key for the transfer of pollutants to higher trophic levels. It has been demonstrated that POPs tend to sorb to phytoplankton rich in organic carbon [Skoglund et al (1996)], which supports the idea that the cycling of POPs and the cycling of organic carbon are related. Though, the interactions between partitioning to organic matter and pollutant fate have received some attention on a local and global scale, however, the field is still far from being understood in a comprehensive manner.

1.3 Modelling approaches

Numerical models are useful tools to study the environmental distribution and fate of pollutants, as field measurements are costly, difficult to conduct (pollutants are only present in low concentrations in the environment), and have usually only a limited temporal and spatial coverage. Numerical models have the added benefit of being suitable for testing hypotheses on underlying processes. The quality of model results depend on the quality of process parameterisations, input parameters, i.e. physical and chemical properties of the substances, quality of spatial and temporal emission patterns and compartmental split upon entry. Principally, two types of models are used: multimedia mass-balance box models (MBMs) and transport models . In multimedia mass-balance box models the environment is considered to consist of a series of 'well-mixed' boxes (e.g. soil, atmosphere, ocean) [Mackay (1991)]. The exchange processes between the boxes (compartments) are described by mass transfer kinetics or instantaneous equilibria. For each compartment a differential mass balance equation, which equates the inventory change in the compartment with source and sink processes is solved. These models are easy to construct and handle, and the computational effort needed to run them is low. For these reasons they are used in decision making, such as substance screening for risk assessment. Furthermore they are also applied to mimic substance fate, and to test individual process parametrisations. One main disadvantage of MBM models for studying global fate of pollutants is their low spatial resolution. In contrast, transport models in addition to inventory changes within the compartments and exchange processes, capture dynamic features of the environmental media, i.e. fluid transport and waves dynamics in atmosphere and/or ocean with relatively high spatial and temporal resolutions. The high process and spatial resolutions make these models computationally expensive, and difficult to handle. The main advantage of these models however is, that they are geo-referenced, which allows a better evaluation of model results by comparison with observations, than for MBMs. Environmental conditions, such as precipitation, wind, soil hydrological status, clouds etc. are not represented as temporal means, but fluctuate in time following statistics similar to real world conditions. These models are the most sophisticated tools available and best suited to study the complex interrelations of processes spatially resolved.

1.4 Objectives and outline of the thesis

As mentioned above, the role of the ocean in global pollutant cycling, specifically the impacts of suspended organic matter on its storage capacity and on long-range transport, need further investigation. One objective of this study is to examine the fate of persistent organic contaminants in the marine environment based on the findings of Guglielmo (2008), and in particular to assess the importance of marine suspended matter. For this purpose simulations using the global multicompartment chemistry-model MPI-MCTM were conducted. First, the representation of continental shelves and sinking particulate matter in the ocean compartment was refined. The model was then used to study the impact of the different representations of marine organic matter on the environmental fate of DDT. The study focusses on the following questions:

- How strong is the impact of different representations of suspended organic matter on the distribution of biogenic tracers (phytoplankton, zooplankton, organic carbon)? Do the representations indeed lead to more realistic distributions of these tracers?
- What is the impact of the different representations on the contribution of organic matter to transport of pollutants into the deep ocean?
- Do the representations lead to a change in long-range transport of DDT and if so, which features of suspended organic matter are relevant for long-range transport?

Furthermore, in Chapter 1 the model is used to examine the relative importance of sea surface temperature, wind speed and pollutant concentration on the volatilisation of DDT. The main questions of this section are:

- What is the spatial distribution of the relative significance of sea surface temperature, wind speed, and pollutant concentration for the variability of the volatilisation of DDT? Are there temporal and spatial regimes in which one of the parameters is more important than the others?

- How important is spatially resolved volatilisation in comparison with spatial means?

In Chapter 2 studies of the environmental fate of selected organic contaminants are presented. The first of two substances is the lipophilic insectice DDT, which is regulated under the Stockholm Convention. The second substance is an 'emerging substance' of the group of the perfluorocarbons (PFCs), perfluorooctanoic acid (PFOA). This substance acts as a surfactant and differs strongly from DDT in its behaviour in the environment. In these studies the past and present day (DDT 1950-1990, PFOA 1950-2004) distributions of the contaminants are simulated based on historical emissions. In addition to compartmental distribution and mass budget, for PFOA special focus is given to assess the efficiency of transport to the Arctic ocean by oceanic currents. In contrast to DDT, which is semivolatile, it dissociates in water and in this form is not volatile. The model results are compared with observational data, and thereby the performance of the MPI-MCTM is evaluated.

1.5 Current state of the model

The non-steady-state dynamical multicompartment chemistry-transport model (MPI-MCTM) used in this study is based on a coupled atmosphere-ocean general circulation model. It consists of the following submodels: the general atmospheric circulation model ECHAM5 [Roeckner et al (2003)] including a microphysical aerosol model (HAM), the oceanic general circulation model MPIOM [Marsland et al (2003)] with the marine biogeochemistry model HAMOCC5 [Maier-Reimer et al (2005)] embedded, and the surface mass exchange parameterisations [Lammel et al (2001)]. Here only main features are summarised. Details of parameterisations implemented can be found in the literature mentioned above as well as in Semeena (2005) and Guglielmo (2008). A schematic overview over the model components and processes relevant for chemicals modelling is given in Figure 1.1.

The atmosphere model The atmosphere model ECHAM5 is a general circulation model based on the numerical weather prediction model of the European Centre for Medium-Range Weather Forecast (ECMWF) [Roeckner et al (2003)]. The prognostic variables of the model are vorticity, divergence, temperature, and surface pressure. They are represented in the horizontal by truncated series of spherical harmonics. Water vapour, cloud liquid water, cloud ice and trace components are transported with a semi-Lagrangian transport scheme [Lin and Rood (1996)] on a Gaussian grid. ECHAM5 contains a microphysical cloud scheme [Lohmann and Roeckner (1996)] with prognostic equations for cloud liquid water and ice. Cloud cover is predicted with a prognostic-statistical scheme solving equations for the distribution moments of total water [Tompkins (2002)]. Convective clouds and convective transport are based on a mass-flux scheme [Tiedtke (1989), Nordeng (1994)] The dynamic aerosol model HAM (Hamburg Aerosol Model [Stier et al (2005)]) describes the atmospheric aerosol spectrum in terms of log normal distributions for 7

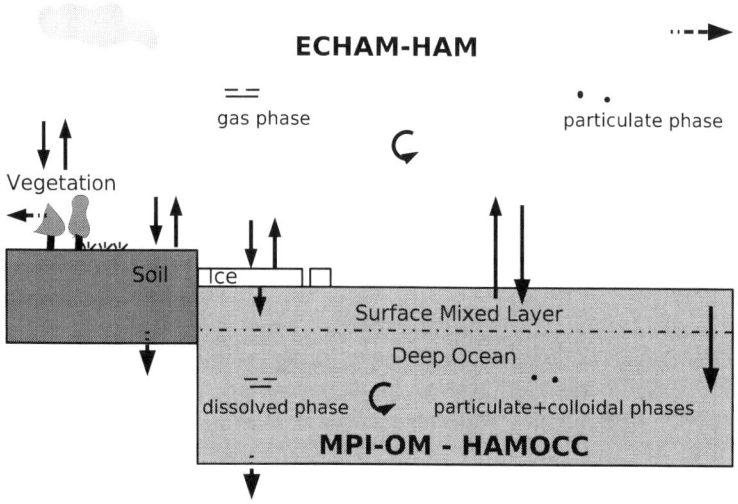

Fig. 1.1: Multicompartmental model world, schematic overview. Dashed arrows denote sinks, solid arrows exchange processes.

aerosol modes. The modes of the aerosol model are composed either of compounds with no or low water-solubility (insoluble mode), or by an internal mixture of insoluble and soluble compounds (soluble mode) [Stier et al (2005)]. The composition of each internally mixed mode can be modified by aerosol dynamics, e.g. coagulation, by thermodynamical processes, e.g. condensation of sulfate on preexisting particles, and by cloud processing. Aerosol number and mass are transported as tracers.

The ocean model MPIOM (Max Planck Institute Ocean and Sea Ice Model) is solving the primitive equations for a hydrostatic Boussinesq fluid with a free surface on a rotating sphere [Marsland et al (2003)]. The vertical discretisation is on z-levels and the bottom topography is resolved with partial grid cells [Wolff et al (1997)]. The spatial arrangement of scalar and vector variables is formulated on a C grid [Arakawa and Lamb (1977)]. The along-isopycnal diffusion follows Redi (1982) and Griffies (1998). Isopycnal tracer mixing by unresolved eddies is parameterised following Gent et al (1995). For the vertical eddy viscosity and diffusion a scheme of Pacanowski and Philander (1981), in which coefficients of eddy mixing are Richardson-number dependent, is applied. In the presence of static instability, convective overturning is parameterised by enhanced vertical diffusion. A slope convection scheme has been included that allows for a better representation of the flow of statically unstable dense water over sills, such as in the Denmark Strait or in the Strait of Gibraltar and off shelves, such as on the Arctic and Antarctic shelves.

The model includes a dynamic thermodynamic sea ice model. The dynamics of sea ice are formulated using viscous-plastic rheology [Hibler (1979)]. The thermodynamics relate changes in sea ice thickness to a balance of radiant, turbulent, and oceanic heat fluxes. The effect of snow accumulation on sea ice is included, along with snow-ice transformation when the snow/ice interface sinks below the sea level because of snow loading. The effect of ice formation and melting is accounted for within the model assuming a sea ice salinity of 5 psu[1].

The marine biogeochemistry model The biogeochemistry model HAMOCC (Hamburg Ocean Carbon Cycle Model) is implemented into the MPIOM physical ocean model [Maier-Reimer et al (2005)]. HAMOCC is designed to address large-scale long term features of the marine carbon cycle. HAMOCC is a (nutrients, phytoplankton, zooplankton, detritus) NPZD model based on the colimitation of nutrients such as phosphorus, nitrogen, and iron (depending on the model version). MPIOM computes advection and diffusion of the biogeochemical tracers as it does for salinity or temperature, except for the omission of isopycnal mixing for biogeochemical tracers. HAMOCC is driven by the same radiation as MPIOM to compute photosynthesis. Biological production is temperature independent assuming that phytoplankton adopts to local conditions. Temperature and salinity provided by the ocean model are used to calculate various transformation rates and constants e.g., for solubility of carbon dioxide.

Substance cycling parameterisations The model considers inter- and intracompartmental mass exchange and conversion processes for the compartments atmosphere, ocean, sea ice, soil and vegetation [Lammel et al (2001), Semeena (2005), Guglielmo (2008)]. Contaminants can be introduced into the model in all compartments, where application and emission can be defined by local conditions, such as daylength and precipitation. Apart from chemical degradation, burial in sediments is considered as final loss to the model environment.

The atmosphere is a three-phase system (gas, particles, cloudwater), with the mass exchange between the phases being controlled by instantaneous equilibria. Degradation is controlled by the hydroxyl and nitrate radical concentrations, provided by HAM. Those 3D fields are prescribed from calculations of the chemistry-transport model MOZART [Stier et al (2005), Horowitz et al (2003)]. For the organic substances studied, partitioning between gas and particles is considered as being determined by absorption into organic matrices, and is predicted from an empirically derived regression [Finizio et al (2007)]. The organic matrices are described as the sum of the aerosol species provided by HAM. The flux of trace substances from the atmosphere to underlying surfaces (vegetation, soil, ocean, and sea ice) includes wet and dry deposition. The dry deposition flux of gaseous substances is calculated using dry deposition velocities calculated according to a resistance scheme [Ganzeveld and Lelieveld (1995)]. Particle dry deposition is a function of aerosol properties, such as particle radius and density. Wet deposition is calculated for stratiform and convective clouds.

Soil and vegetation are represented as two-dimensional compartments with no vertical resolution. The transfer of water within the soil is calculated using a bucket-type

---

[1] PSU (practical salinity units) are a unit of measurement of salinity, i.e. of the total amount of dissolved salts in water. If sea water has a salinity of 5 psu, 5 g of salt are dissolved in 1000 g of water.

approach [Roeckner et al (2003)]. Soil is considered as a three phases system consisting of organic matter, pore water and pore air. Volatilisation from the soil in MCTM is described by an empirically derived parameterisation of volatilisation of pesticides [Smit et al (1997)] following establishment of thermodynamic equilibrium in soil. Vegetation, which covers a temporally varying fraction of the land grid cells, is represented as a surface only. For pollutants the processes application, storage, degradation and (re-)volatilisation are described depending on vegetation characteristics (plant type, phenology) and local conditions. Volatilisation is described using an empirically derived parameterisation [Smit et al (1998)]. Surface layer temperatures define degradation rates in soil and vegetation. Degradation processes in soil and vegetation are represented as first-order processes. In soil and vegetation degradation is assumed to double per 10 K temperature increase.

Processes considered in the ocean compartment are: partitioning to colloidal and particulate matter with the sinking of particulate-bound contaminants, volatilisation to the atmosphere from the dissolved phase, degradation of dissolved compounds, and transport of contaminants due to mixing of water masses by turbulent diffusion and advection. The non-sinking colloidal phases is determined by the sum of phytoplankton, zooplankton and dissolved organic carbon, whereas the sinking phase is determined by detritus. The partitioning is calculated as an instantaneous equilibrium. The flux from the ocean to the atmosphere is parameterised as based on the two-film model [Whitman (1923)] using the fugacity formulation [Mackay (1991)]. Fugacity is a thermodynamic quantity that can be viewed as the escaping tendency of a substance from a phase [Schwarzenbach et al (2003)]. In the two-film model the state of the atmosphere is represented by temperature and wind velocity and the parameterisation follows empiric relationships based on inorganic trace gases [see Schwarzenbach et al (2003)].

Sea ice is represented in the model as a two-dimensional surface covered with a snowpack. Ice advection, rheology and snow cover are calculated from the sea-ice model embedded in MPIOM [Hibler (1979)]. The only source of pollutants for the ice compartment is deposition from the atmosphere. Once pollutants enter the ice compartment they can diffuse into the snow pore space air, dissolve in the interstitial liquid water or adsorb to the ice surface. Together with the sea ice the pollutants undergo advection. Sinks considered for the ice compartment are volatilisation to the atmosphere and release into the ocean with melt water.

Chapter 2
Model development

2.1 Introduction

The role of the ocean as a transport medium, as well as a storage reservoir, for persistent organic pollutants has been widely studied to a lesser extent, but is not fully understood yet. Suspended matter in the ocean, and in particular, particulate organic matter that sinks below the euphotic zone is assumed to play a significant role in the fate of organic pollutants that are present in natural waters and have a high tendency to partition to organic matter [Gustafsson and Gschwend (1997), Dachs et al (2002), Jurado et al (2007)]. Pollutant mass bound to organic matter in the surface ocean is not available for volatilisation. Sinking of organic matter below the mixed layer is one process that transfers pollutants into the deep ocean, from where only upwelling and overturning can bring them back to upper oceanic layers. The relevance of partitioning to organic matter in the ocean has hardly been studied at the global scale. Measurements revealed high fractions of pollutants bound to phytoplankton and particulate matter in the ocean [c.f. Tanabe and Tatsukawa (1983), Dachs et al (1996)]. Box model and laboratory studies suggested the importance of partitioning of pollutants on phytoplankton for air-sea exchange, vertical transport and on the ocean's capacity to act as a reservoir for persistent organic pollutants. The work of Dachs et al (1999) and Dachs et al (2002) indicates the role of the phytoplankton uptake and settling of particulate matter as drivers of the oceanic sink of POPs such as polychlorinated biphenyls (PCBs) and dibenzo-p-dioxins and -furans (PCDD/Fs). Also zooplankton and its fecal pellets have been reported to be an important pathway for organic contaminants [Wakeham et al (1980), Fowler and Knauer (1986) Dachs et al (1996)]. It was furthermore suggested that the export to deep ocean affects the atmospheric residence time of POPs [Scheringer et al (2004), Jurado (2006)].

Addressing a global multicompartmental chemistry-transport model including a circulation model and a marine ecosystem model bears the advantage of including all relevant processes for the study of the effect of marine organic carbon on long-range transport with sufficient resolution. This kind of global multicompartmental chemistry-transport model was used by Guglielmo (2008) to study the role of the ocean as a transport medium as well as the potential of marine organic matter to transfer pollutants into the deep ocean. In her model experiments half of the DDT reaching the ocean by atmospheric deposition had been transported below the mixed layer within approximately five years. Substance transfer to the deep ocean is believed to be underestimated in the model setup used by Guglielmo (2008), and the design of the model experiments makes it impossible to isolate the contribution of organic matter to long-range transport. The underestimation is due to the biogeochemistry model HAMOCC which is designed to simulate large-scale mean processes, and does not resolve the high gradient of organic matter observed in continental shelf zones. Coastal zones, especially continental shelves are one of the most biological diverse and productive ecosystems in the world. They are by far more active and show a higher diversity than the open ocean due to a high nutrient supply from the continent. As a high amount of DDT was predicted to be bound to organic matter, which is locally up to 30 % of the DDT mass present in seawater [Guglielmo (2008)], a better representation of continental shelves in the global model is likely to be key for comprehensive modelling of such substances in the marine environment, and hence for long-range transport. Optical sensors that measure ocean colour from space can be used to produce global maps of chlorophyll-a concentrations. Although these have a number of shortcomings they provide (the best available) information about the oceanic state on a global scale with a high temporal and spatial resolution. Therefore, chlorophyll-a concentrations derived from the MERIS instrument of the European Space Agency (ESA) are used to artificially impose a high gradient in chlorophyll concentration on the ocean margins in the model.

In the current study, also processes related to sinking of organic matter in the ocean are refined. Guglielmo (2008) used HAMOCC5 with homogeneous sinking speeds of particulate organic carbon of 5 m/d. The current version of the model includes the option of simulating aggregation of marine snow, thereby allocating temporally and spatially varying sinking speeds to the sinking fraction of organic matter.

In the following chapter model refinements are described and compared with the setup used by Guglielmo (2008). The focus is given on the represention of marine organic matter. In a sensitivity study the impact of organic matter on long-range transport is explored. Additionally, a study is included that clarifies the relative importance of sea surface temperature, wind speed, and pollutant concentration for volatilisation of DDT from the ocean.

2.2 Model refinements

The refinements made to the entire model setup include a higher model resolution, the implementation of most recent ECHAM, MPIOM and HAMOCC versions, the usage of assimilated satellite data for surface phytoplankton distribution, and the usage of a more realistic description of sinking organic matter in the ocean.

Table 2.1 gives an overview of the new features in the current model setup in comparison to the one used by Guglielmo (2008). Detailed descriptions of the individual refinements are given in the following paragraphs.

Table 2.1: Overview over model refinements.

Component	Feature	Previous version[a]	Current version
ECHAM	horizontal resolution	T42	T63
	grid points	128x64	192x96
	grid spacing	300 km	225 km
	time step	1800 s	1200 s
MPIOM	horizontal resolution	T42	GR15
	grid points	128x64	254x220
	grid spacing	300 km	12-170 km
	vertical resolution	13 levels	40 levels
	time step	1800 s	4320 s
HAMOCC	sinking speed of POC	5 m/d	size dependent sinking
	continental shelves	no special treatment	phytoplankton concentration corrected with MERIS chla2 products
GCM coupler	software	ECHAM/MPIOM	OASIS3
	method	gridpoint to gridpoint	interpolation, regridding
	frequency	3 hours	24 hours

[a] Guglielmo (2008)

Updated model versions The atmosphere and ocean models were updated to the recent model releases ECHAM5.3.02 and MPIOM-1.2.2. The update of the biogeochemistry model from the beta version of HAMOCC5 used by Guglielmo (2008) to HAMOCC5.1 [Maier-Reimer et al (2005)] introduced major changes in the model.

These updates include additional processes and tracers. Principally the computation of the biogeochemistry is based on colimitation of nutrients. The beta version used by Guglielmo (2008) is based on the colimitation of phosphorus and nitrogen, whereas the current HAMOCC5.1 additionally accounts for iron limitation. Furthermore HAMOCC5.1 includes nitrous oxide, and dust. HAMOCC5.1 also simulates formation of calcium carbonate and opaline shells by phytoplankton thereby differentiating between two phytoplankton groups. Both model setups differ in their tuning parameters, such as the remineralisation and mortality rates of the biogenic tracers.

The previous model setup included an identical horizontal resolution for ocean and atmosphere, and online coupling was implemented in the ECHAM submodel with an exchange rate of 3 hours. In the current model setup atmosphere and ocean are coupled online using the Ocean-Atmosphere-Sea Ice-Soil (OASIS) coupler [Valcke et al (2004)] with a coupling time step of one day. OASIS passes boundary conditions of the two submodels MPIOM and ECHAM5 (sea surface temperature, sea ice variables, heat, freshwater and momentum fluxes, downward radiation and 10 m wind speed), and volatilisation and deposition fluxes for the chemical transport model between the two models.

Model resolution Guglielmo (2008) used a setup in which the atmosphere model operated on a T42 spectral grid and resolved 19 pressure levels up to 10 hPa. In the ocean a spatial resolution approximating spectral truncation T42 was used, vertically resolving 13 levels of varying width, with a higher resolution in the uppermost levels. Photosynthesis in HAMOCC took place in the level 1 to 3, which corresponds approximately to 60 m depth. To obtain a better representation of continental shelves a higher resolution was chosen for the current model setup. The ocean and atmosphere models operate on different grids. The horizontal resolution of the atmosphere model is T63 in spectral space with a corresponding resolution of $1.9°$x$1.9°$ on a Gaussian grid. In the ocean an orthogonal curvilinear grid with the North Pole shifted to Greenland and the South Pole moved towards the centre of the Antarctic continent is used. This approach removes the numerical singularity associated with the convergence of meridians at the geographical North Pole and produces higher resolution of approximately 12 km in the deep water formation regions in the Greenland Sea, Labrador Sea, and Weddell Sea [Marsland et al (2003)]. The coarsest resolution is approximately 170 km in the tropical Pacific. The nominal resolution of the ocean grid is 1.5 °, referred to as GR15 in the following. In the vertical 40 layers are resolved, 8 layers within the upper 90 m, and 20 in the upper 600 m. Photosynthesis takes place in the first 8 model layers. In the following these first 90 m are referred to as the euphotic zone of the model.

Sinking of particulate organic carbon In general, fecal pellets and the so-called 'marine snow' are thought to be of importance for sedimentation in the ocean. Fecal pellets usually play a role in sedimentation when large zooplankton ($> 100 \ \mu$m) is present, whereas marine snow is of importance whenever particles are abundant [Cherry et al (1978), Fowler and Knauer (1986)]. Marine snow aggregates arise from coagulation of smaller, suspended particles which after collision stick together. Particles collide because of their relative movement with the water (shear) or through it (differential settling). Marine snow may consist of very different kinds of particles, such as diatoms, smaller phytoplankton, fecal pellets, detritus, and other types of particles [Cherry et al (1978), Asper (1987), Alldredge and Gotschalk (1989)]. The composition of marine snow may change with time. Marine snow particles can also break apart due to shear; and bacterial degradation can lead to their dissolution. Hence the particle size distribution is a function of several variables including source and nature of the particles, and physical or biological processes of aggregation. The distribution can often be approximated by a power law function [McCave (1984)]. Depending on their diameter and density particles sink with velocities varying from centimetres to several hundred meters per day [Carder et al (1982), Azetsu-Scott and Johnson (1992)]. Large particle tend to sink faster than smaller ones. But, as marine snow aggregates are fractal objects, with mass growing slower than diameter, the mass-diameter relationship also follows a non-trivial function and has to be considered in the relation of diameter and sinking speed.

The biogeochemical model HAMOCC, as used by Guglielmo (2008) considers only one sinking phase, detritus, which sinks with a homogeneous velocity of 5 m/d. Phytoplankton, zooplankton, and dissolved organic carbon are passively transported by advection and diffusion. An implementation of a model for aggregation of marine snow in HAMOCC [Maier-Reimer et al (2005)], developed by I. Kriest [Kriest (2002)], allows assigning sinking speeds to particles based on a particle size distribution of marine snow, whereby the latter is assumed to consist of phytoplankton and detritus. The particle size distribution and its variations caused by aggregation and sinking are calculated explicitly. Here the approach is described briefly, for details refer to [Kriest (2002), Maier-Reimer et al (2005)]. The approach assumes that the relations between particle diameter and phosphorus, and between the sinking speed of individual particles and their diameter follow a power law with fixed slopes. Also the size distribution of particles is approximated by a power law function with a temporally varying exponent depending on number and mass of the aggregates. Number of particles and mass of particles are considered separately. Number of particles changes due to photosynthesis, exudation, grazing, remineralisation, and zooplankton mortality. Photosynthesis creates particulate organic material, whereas exudation, grazing, and remineralisation destroys mass. Dead zooplankton comes with its own flat distribution and always adds large particles. The number of particles also changes due to aggregation and sinking of particles, which preferentially removes the larger particles. Aggregation depends on the particle abundance, their size distribution, the rate of turbulent shear and the difference in particle sinking speeds, as well as their stickiness. The stickiness depends on the abundance of dead and living diatoms, diagnosed from the amount of opal, and increases with opal abundance. The fact that also the living cells sink when aggregation is simulated affects the representation of opal and calcium carbonate. It differs form the default setup, when only the shells (opal and calcium carbonate) that are part of detritus (i.e., that have been grazed and egested as fecal pellets or are contained in dead zooplankton and phytoplankton) had been accounted for. The average sinking velocity calculated from the particle size distributions and the diameter-sinking speed relation defines sinking of detritus, phytoplankton, opal, dust and calcium carbonate. As in the default case only pollutants in the particulate phase are sinking. The colloidal phase is affected by sinking indirectly only, namely via sinking of living phytoplankton that changes the vertical distribution of organic matter.

Continental shelf zones The high gradient of organic matter on the ocean margins, not resolved by HAMOCC, is introduced into the model by assimilating chlorophyll-a concentrations observed from satellite in shallow waters. Chlorophyll-a is the main pigment of many phytoplankton species [Jeffrey et al (1997)] and is used here directly as a proxy for phytoplankton concentration. Phytoplankton photosynthetic pigments consists of chlorophyll-a, the accessory pigments chlorophyll b and c, and the photosynthetic carotenoids [Martin (2004)]. The approach of taking it as a measure for phytoplankton abundance is justified, because chlorophyll-a is the only photosynthetic pigment that occurs in all phytoplankton [Martin (2004)].

For this very purpose the data product chlorophyll concentration is derived from optical sensors. Over the ocean these passive remote sensing devices measure light reflected by the Earth's surface consisting of contributions of the water body and radiation scattered in the atmosphere. Only reflected light from the ocean carries information about its constituents, hence the signal needs to be corrected for atmospheric contributions. Over open ocean waters atmospheric corrections are based on the assumption of no light penetration in and backscattering by the ocean in the near infrared (NIR) [Gordon and Morel (1983)]. These waters, in which the optical properties are only defined by chlorophyll, are defined as Case 1 waters [Morel and Prieur (1977)]. Remote sensing algorithms meant to derive chlorophyll concentrations from ocean colour in Case 1 waters are based on colour ratios of 2 channels in the NIR (blue over green [Gordon and Morel (1983)]), which are related empirically to the chlorophyll-a concentration. In coastal waters the atmospheric signal in the NIR can not be separated, because of the strong diffuse reflection which is caused by backscattering of particles suspended in the water. These optically complex waters, in which besides phytoplankton pigments also yellow substance (or coloured dissolved organic matter CDOM) and suspended particulate matter have a strong impact on light attenuation, are defined as optical Case 2 waters. Case 2 water algorithms need to include spectral information from all channels to derive suspended matter contents of the ocean.

The chlorophyll-a concentrations for Case 2 waters used in this study were derived from the European Space Agency's Medium Resolution Imaging Spectrometer (MERIS). MERIS is an optical sensor flying on the ENVISAT-1 satellite and was designed especially for ocean and coastal water remote sensing [Rast (1999)]. It has a spatial resolution of 300 m, and a revisit period of 1-3 days. 9 of its 15 spectral bands are in the range 412-708 nm. One band measures the fluorescence of phytoplankton and it has a high signal to noise ratio above dark targets as found over the open ocean in the NIR. The operational MERIS Case 2 water algorithm to derive chlorophyll-a concentrations [Doerffer and Schiller (2007)] is based on a neural network (NN) [Doerffer and Schiller (2000)] which uses the ground reflectances measured in 8 bands, solar and viewing zenith, and azimuth difference to calculate three inherent [1] optical properties. These are: scattering of all particles, absorption of phytoplankton pigments and absorption of yellow substance and the bleached fraction of SPM. The NN is trained by a biooptical model, which in turn is based on a large data set of inherent optical properties. These data have been collected on cruises in the North Sea, Baltic Sea Mediterranean and North Atlantic. The environmental conditions for the model to be valid are, infinite deep water (no bottom reflectance), vertically homogeneous distribution of all water constituents, a rough sea surface (i.e. approximately 3 m/s wind speed). The Level-3 product, i.e. monthly mean global maps, was produced from the L2 product within the GLOBCOLOUR project [ACRI-ST et al (2007)] and were taken from their website (http://www.globcolour.info/). Monthly

---

[1] Inherent optical properties are independent of changes in the radiance distribution and depend only on the substances in the water [Preisendorfer (1961)].

mean data from 2002-2006 were compiled into a set of multiyear means for each month of the year.

Assimilation At the end of each month phytoplankton concentrations in the surface level were adjusted according to the MERIS satellite chlorophyll-a concentrations of the corresponding month. The data assimilation is only done for shallow waters, with a water depth lower than 250 m (see Figure 2.1).

For this purpose Case 2 water chlorophyll-a concentrations from MERIS were transformed into phytoplankton concentrations using the constant carbon-to-chlorophyll ratio, $R_{C:Chl}=60$, used in HAMOCC [Maier-Reimer et al (2005)].

The modified phytoplankton concentration, phy_n is derived from

$$phy_n = phy_{n-1} + a \cdot (phy_{n-1,MERIS} - phy_n) \tag{2.1}$$

where n is the number of month, and $phy_{n-1,MERIS}$ the phytoplankton concentration derived from the MERIS chlorophyll-a data. The difference is weighted by a, given as:

$$a = V_v \cdot (1 - V_{int} \cdot V_{MERIS}) \tag{2.2}$$

where V_{int} is a factor that takes into account the interannual variability of the phytoplankton mass. It is derived from MERIS data by calculating the relative standard error ($\frac{\sigma}{N\text{mean}}$, N - number of years) of the interannual mean. The factor V_v corrects for lacking variability within one month in the MERIS data. It represents the ratio between the phytoplankton concentration at the last time step of a month in HAMOCC and the corresponding monthly mean. V_{MERIS} equals the error bar of the MERIS chlorophyll-a product (0.713 for Case 2 waters [ACRI-ST et al (2007)]).

Although the adjustment of model phytoplankton concentrations takes places every month, not all shallow water locations are affected by it every month. Since MERIS is an optical sensor, light availability limits its ability to measure ocean colour. Hence solar angle and clouds determine, whether assimilation is possible or not. The number of months in which is assimilation is possible is higher for locations close to the equator than locations at higher latitudes (Figure 2.1b)).

2.3 Impact of the horizontal resolution on the representation of continental shelves

2.3.1 Introduction

Bottom topography plays a major role in determining the flow field in the ocean. Currents, tides, mixing, and upwelling are influenced by topography. In shallow water regions the impact of topography is especially pronounced. In HAMOCC the topography has a strong impact on the flux of organic carbon into the sediment. Regarding pollutant cycling, differences in the bottom topography translate into different amounts of a pollutant permanently stored or degraded in sediment, because resuspension from the sediments is not yet included in the model.

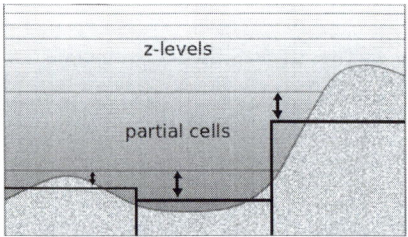 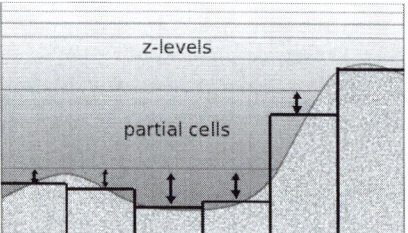

Fig. 2.2: Vertical z-levels with partial cells in the bottom level. Bottom topography is shown for arbitrary coarse and high horizontal resolutions

The bottom bathymetry of MPIOM is derived from interpolating the 5-minute grid-ded ETOPO5 data set [NOAA (1988)] onto the model grid. The latter combines various sources of land elevation data with a digital bathymetric data base produced by the U.S. Naval Oceanographic Office from hand-drawn contour charts. In MPIOM only the horizontal model resolution defines the accuracy of the sea floor topography [Marsland et al (2003)], although it is vertically discretizised on z-levels. In a z-level discretisation large distances between the levels in the deep ocean influence the res-olution of the bottom topography. But, MPIOM includes partial vertical cells, i.e. at each point in the horizontal grid the deepest wet cell has a uniform thickness that is adjusted to resolve the discretised bathymetry (Figure 2.2). Therefore, increased horizontal resolution offers a better representation of bottom topography features. To assess the improvements in the representation of continental shelves due to the enhanced model resolution, the area covered by selected continental shelves and their mean depths are compared for the two horizontal resolutions T42 and GR15.

2.3.2 Methodology

The GR15 and T42 topography originate from the ETOPO5 [NOAA (1988)] data set interpolated onto the model grid. Hereby, specific topographic features, such as the important conduits of overflows and throughflows, were adjusted to observed depths [Marsland et al (2003)].

Depths and area of 16 selected continental shelfs for MPIOM in two different hori-zontal resolutions T42 and GR15 were compared among each other and with global seafloor 2-minute gridded topography ETOPO2, which represents the truth in this study. The latter is the 2001 version of the topography data collection of NOAA [NOAA (2001)]. Similar to ETOPO5 it was constructed from a variety of sources, but it mainly consists of data from satellite altimetry of the sea surface.

The comparison was conducted by first defining regional boxes containing continental shelfs zones (Figure 2.3). Within the individual boxes the area of all gridcells in which the water depth is lower than 250 m is calculated and summed up. In the literature one often finds 150 m as the maximum depth of continental shelfs. The threshold of 250 m is an arbitrary value, used to ensure that the number of grids cells is reasonably large also in the coarse resolution, T42. In GR15 the non-rectangular shape of the grid cells makes the computation of their area difficult. A simple approximation method was used: For every grid cell the maximum and minimum size was determined from the minimal and maximal distance between the grid corners assuming quadrature. From these minimum and maximum areas a mean value is derived, which is used in the comparison. Minmum and maximum values can be found in the Appendix.

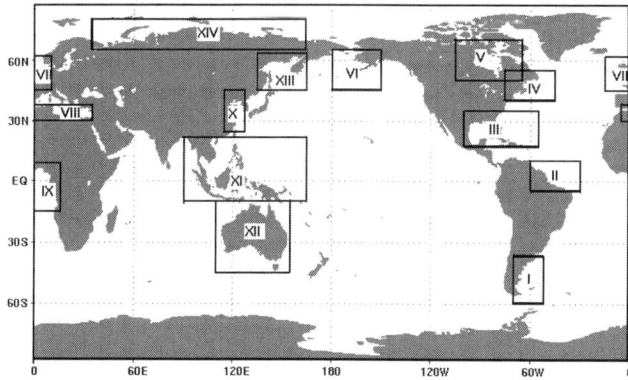

Fig. 2.3: Position of the continental shelf regions.

2.3.3 Results and discussion

Generally, the area of continental shelves in the model bathymetries is lower than the area estimated from the ETOPO2 data set (Table 2.2). The total continental shelf area of the considered regions is 17 800 926 km^2 in ETOPO2, 15 532 692 km^2 in GR15, and 9 016 847 km^2 in T42. In the boxes IV, V, VI, and VII, in the vicinity of the model North Pole, the grid cells in GR15 are strongly asymmetric. The difference between minimum and maximum area derived from minimal and maximal distances of the grid corners is especially large in these boxes (see Appendix, Table A.1). Therefore, the area of the cells includes uncertainties. In most regions the area is lower in T42 than in GR15. The Amazon Continental Shelf and the shelves in Central Africa, and in the Southern Mediterranean Sea are not resolved in T42.

The comparison of ETOPO2, GR15, and T42 mean water depths show, that in the

Table 2.2: Area [km^2] and mean depth [m] of the continental shelf (CS) zones in the boxes shown in Figure 2.3.

Box	Shelf zone	Area [km^2] ETOPO2	GR15	T42	Mean depth [m] ETOPO2	GR15	T42
I	Patagonian CS	1 010 159	1 074 278	926 414	95.75	91.18	99.46
II	Amazon CS	508 084	304 493	-	44.28	51.00	-
III	Gulf of Mexico	931 481	728 741	256 641	45.52	72.29	98.00
IV	Gulf of St.Lawrence	953 657	879 407	574 729	101.20	123.70	149.87
V	Hudson Bay	1 276 583	967 622	656 059	88.85	110.23	112.00
VI	Gulf of Alaska CS	1 082 189	869 535	614 846	70.43	73.16	55.9
VII	European CS	1 155 507	925 454	305 906	81.17	75.65	95.00
VIII	Mediterranean Sea	291 076	257 251	-	70.58	125.63	-
IX	Central Africa	163 084	156 585	-	64.91	141.22	-
X	East China Sea CS	896 484	934 444	766 097	60.10	65.42	85.67
XI	Sunda Shelf	4 180 507	3 281 129	2 290 955	51.26	60.27	70.50
XII	Australia	2 237 099	2 248 439	685 872	56.95	69.20	81.50
XIII	Sea of Okhotsk	635 779	510 527	437 143	104.44	149.31	107.29
XIV	CS of Russia	2 479 237	2 551 372	1 502 185	77.41	97.59	116.73

model topographies the water depths is higher in most continental shelf regions. The differences in the mean depths between the topographies are mainly driven by the chosen threshold of 250 m. Especially in regions with only few grid cells, such as the Gulf of Mexico, a single cell can determine the average. In the Gulf of Alaska Continental Shelf is the shelf is in T42 only 56 m deep, whereas GR15 and ETOPO2 show that the average is approximately 70 m. Shelf areas in which the water depths is lower in GR15 than in ETOPO2 are the Amazon Continental Shelf and the European Continental Shelf. The Amazon Continental Shelf is one of the smallest shelf zones and is situated in the area of GR15, where the horizontal resolution is coarse [Marsland et al (2003)]. Thus, also here the mean water depth is dervied from only a a few grid cells, hence from a deficient representation of the bottom topography (Figure 2.2). Strongest discrepancies between ETOPO2 and T42 occur in the Gulf of St. Lawrence and the Continental Shelf of Russia, where the shelf in T42 is approximately 50 % deeper than the ETOPO2 shelf. In both regions the area of the continental shelf is only 60 % of the size of the shelf in the ETOPO2 data set.

2.3.4 Summary

In the higher resolution (GR15) the representation of continental shelves is much better than in the coarse resolution (T42), both in terms of area, as well as in terms of water depth. T42 resolves only 51% of the total shelf area found in ETOPO2. Most of the shelves resolved by the model are deeper than the ones in ETOPO2, whith an average deviation of 36 %. The continental shelves in the Mediterrainian Sea, in Central Africa and the Amazon Continental Shelf are not represented in T42.

GR15 resolves approximately 87 % of the continental shelf area. Most continental shelves are deeper than the ones in ETOPO2. The average deviation of the water depths is 32 %.

2.4 Sensitivity of long-range transport of dichlorodiphenyltrichloroethane to marine organic matter

2.4.1 Introduction

Numerical models produce results dependent on many input parameters and process parametrisations, which with regard to pollutants cycling, include degradation rates, partitioning coefficients, diffusion coefficients and all parameters which affect the environmental state of the model. In general, to understand the sensitivity of the model results to a subset of these parameters, the model results are usually compared to observations and a controlling parameter is modified to bring the model results into an estimated error of the thest data set, usually observations. Introducing a variable as a control parameter implies the assumption that the model results are indeed sensitive to that variable and that it is important for reproducing observations. In this study the concentration of organic matter in the ocean represents the controlling parameter for long-range transport of the insecticide DDT. As it is discussed in Chapter 3.1 the uncertainty of input parameters and processes not resolved in the model or insufficiently validated for semivolatile organic compounds are the main reason for discrepancies between model results and observations of DDT. Therefore, the sensitivity study performed does not aim to bring model results into an error estimate of observed DDT concentrations, but rather aims to empirically understand the impact of a set of parameterization for marine organic matter within the model world on long-range transport of DDT to determine the general importance of organic matter rather than to decide which of the parameterisations makes model results more realistic. The premise of the sensitivity study is that organic matter in the ocean is controlling long-range transport by reducing the flux of DDT from the ocean to the atmosphere and enhancing transport to the deep ocean. DDT is a multicompartment substance, which was shown to be transported mainly in the atmosphere [Semeena and Lammel (2005),Guglielmo (2008)]. A reduced volatilisation from the ocean surface, hence a larger fraction of DDT residing the ocean, would presumably lead to reduced long-range transport of DDT. As mentioned above, the results of Guglielmo (2008) indicate that within a relatively short period of time a high fraction of DDT in the ocean sank below the mixed layer. This amount is believed to be underestimated due to processes and characteristics of marine organic matter not resolved in that model setup. For the sensitivity study two experiment that differ only in their representation of marine organic matter were performed. Both experiments attempt to cope with deficiencies of the biogeochemical model HAMOCC as used in Guglielmo (2008). One experiment simulates aggregation of marine snow and related spatially varying sinking velocity. The second experiment uses a spatially and temporally homogeneous sinking velocity of 5 m/d, but includes satellite data assimilation in continental shelf zones. The aim is to conclude from differing compartmental and spatial distribution of the pollutant at the end of the simulation on the significance of marine organic matter for long-range transport of DDT and other substances with similar properties.

2.4.2 Experimental setup

The three dimensional multicompartment chemistry-transport model, MPI-MCTM [Lammel et al (2001), Semeena and Lammel (2003), Guglielmo (2008)], was run for 10 years in two different experiments, one including the aggregation module (AGG), the other one including satellite data assimilation (SAT). The horizontal resolution of the atmosphere model is T63 in spectral space with a corresponding resolution of $1.9°$x$1.9°$ on a Gaussian grid. In the ocean a curvilinear coordinate system with $1.5°$ nominal resolution (GR15, Marsland et al (2003)) is used (details in section 2.2). With regard to organic mater, the two experiments differ in the tuning of the biogeochemical model HAMOCC. This tuning represents parameters affecting remineralisation of detritus, phytoplankton mortality below the euphotic zone and dissolution of calcium carbonate and opal in the total water column (Table 2.3). Both experiments were started from the same initial (restart) conditions, and phyto-

Table 2.3: Parameters used for the discription of dissolution and remineralisation in HAMOCC5 in the aggregation experiment (AGG) and the satellite data assimilation experiment (SAT).

Parameter	AGG	SAT
Detritus remineralization rate $[d^{-1}]$	0.050	0.025
Phytoplankton mortality rate $[d^{-1}]$	0.200	0.100
Opal dissolution rate $[d^{-1}]$	0.030	0.010
Calcium carbonate dissolution rate $[d^{-1}]$	0.200	0.075

plankton shading is not considered to affect the circulation models by modifying the heat budget of the ocean or water leaving radiance, hence both experiments follow the same physical trajectory.

In the experiments the distribution of dichlorodiphenyltrichlorethane (DDT) was simulated based on 1980 applications. Its low water solubility, together with its high lipid solubility (octanol-water partitioning coefficient, $K_{ow} = 1.55 \cdot 10^6$) allows the substance to partition to organic matter in the ocean, and hence satisfying prerequisites to study the impact of organic matter on global pollutant cycling. Details on substance properties and the origin of the applications can be found in Chapter 3.1. In total 7665 t of DDT are applied annually to soil (20%) and vegetation (80%) following the spatial distribution of the 1980 applications (Figure (2.4)). The amount of substance lost directly to air during application was assumed to be zero.

2.4.3 Results and discussion

Impacts of satellite data assimilation on suspended matter evolution A priori, changing only the phytoplankton concentration by satellite data assimilation in the surface layer at the end of each month has manifold implications on the evolution of the biogeochemistry in HAMOCC: An increase in phytoplankton concentration implies more phytoplankton is available for photosynthesis, hence phytoplankton growth is increased. At the same time radiation available for photosynthesis is reduced due to enhanced shading by phytoplankton. This affects growth in the surface level, as well as growth in the level below. Phytoplankton growth takes up nutrients inducing a decrease of the phosphorus concentration. Part of the grown phytoplankton is immediately grazed by zooplankton, which implies that the higher phytoplankton concentration from satellite assimilation leads to an increase of zooplankton grazing. A fraction of grazed phytoplankton is ingested by zooplankton, resulting in an enhanced zooplankton concentration, the remaining fraction is egested as fecal pellets, adding to the detritus pool. Due to the enhanced phytoplankton concentration a higher abundance of colloidal and particulate organic phases is expected after the assimilation in the surface level of the ocean. The satellite data assimilation however is designed to allow enhancement and and reduction of the phytoplankton concentration, preventing overshoots of the biogeochemistry.

In the Australian Shelf at 140°E 10°S satellite data assimilation takes place almost throughout all of the year (Figure 2.5 and Figure 2.1). Except for September the adjustment always increases the phytoplankton concentration abruptly. The increase is especially pronounced in June. The euphotic zone mean values of zooplankton, detritus and dissolved organic carbon concentration (Figure 2.5) do not reveal any abrupt response to the assimilation of phytoplankton. Photosynthesis however, is enhanced immediately due to the increase of phytoplankton, leading to decrease of phosphate which is used for growth of the newly added phytoplankton mass. The reactions of zooplankton and detritus, for example in June, are evident with a time lag, indicating that a large part of the imposed phytoplankton mass is adding to the detritus pool and a minor one is grazed by zooplankton triggering its growth. In September, when assimilation leads to a decrease of the phytoplankton mass impacts are much smaller, due to the smaller difference between the imposed mass and the mass produced by the biogeochemical. Still, a decrease in detritus and zooplankton concentration and a small increase in the phosphorus mass can be observed. At 170°E 65°N assimilation takes place less often throughout the year. In June the impact of the assimilation is especially pronounced, when phytoplankton concentration was decreasing after having reached a maximum in the second half of May, and is abruptly jumping to a value even higher than the former maximum. The phytoplankton is partly consumed by grazing, and dies. This way the dissolved organic

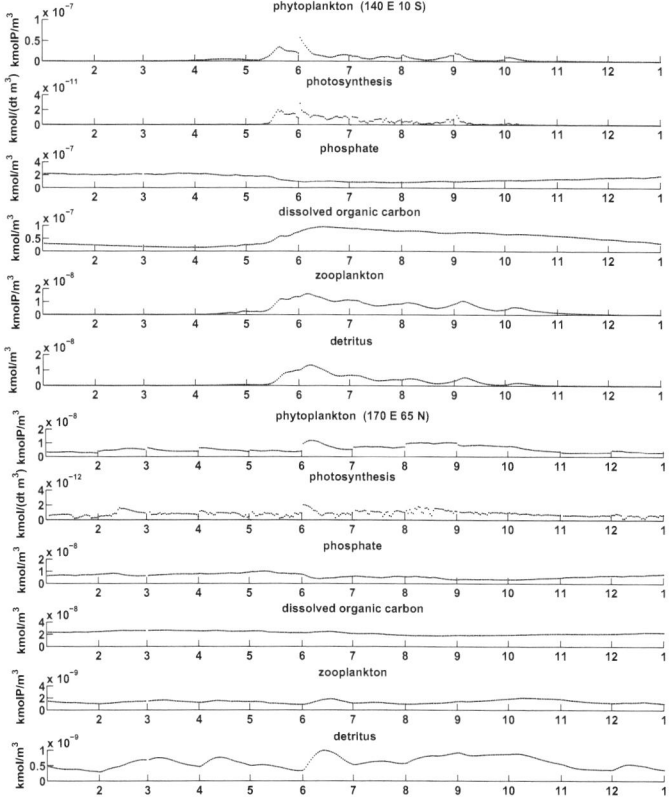

Fig. 2.5: Timeseries of daily phytoplankton, zooplankton, dissolved organic carbon, detritus, and phosphorus concentration, and photosyntesis over one model year at two location the shelf seas of the Pacific Ocean, 170 E 65 N and 140 E 10 S.

carbon, zooplankton and detritus concentrations get enhanced. Already within one week all phytoplankton mass is recycled and the concentrations reaches a value even lower than before the assimilation. Also at $140\,^\circ$E $10\,^\circ$S, after the strong increase in June, elevation of zooplankton and detritus concentrations are observed.

The sign of the satellite data assimilation varies with time and location (Figure 2.6). In the Russian continental shelf, close to the coast in winter and spring the phytoplankton concentrations in the model are lower than the concentrations derived from the MERIS data. Further offshore in winter MERIS phytoplankton concentration on the other hand are lower, thus leading to a decrease of the modelled concentration when assimilation takes place. In comparison to the aggregation experiment the phytoplankton mass is enhanced in continental shelf zones in the assimilation experiment (Figure 2.7). Although the assimilation produces much higher concentrations of phytoplankton than HAMOCC5 does without assimilation, the results are still far from representing the observations: Without assimilation almost all continental shelf locations show a concentration that is underestimated by several orders of magnitude. After the assimilation still the majority of the location show a concentration much lower than the satellite data.

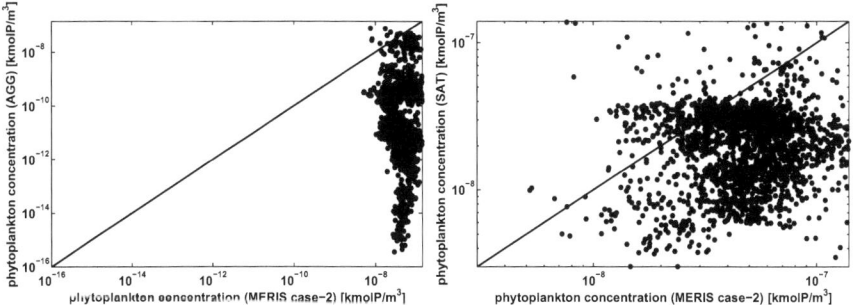

Fig. 2.7: Mean phytoplankton concentration in shelf zones, AGG versus MERIS Case 2 (left) and SAT versus MERIS Case 2 (right).

Aggregation of marine snow and detritus sinking The particle size distribution is modified by two processes, aggregation and sedimentation. Sinking preferably removes large particles leading to an increase of the exponent. Aggregation creates large particles, and affects the number, but not the mass. The simulated mean number of marine snow aggregates follows closely the amount of mass available for aggregation defined by the detritus and phytoplankton concentrations (Figure 2.8). The mean number of cells per aggregate, n, is derived from:

$$n = \frac{m_{snow}}{n_{agg} * m_{min}} \tag{2.3}$$

with the number of aggregates, n_{agg}, the available mass of marine snow, m_{snow}, and the mass of the smallest particle, m_{min}. This variable is an estimate for the mean size of the aggregates. Large slopes of the particle size distributions correspond with small mean aggregate sizes. Due to the compensation impact of sedimentation and aggregation a high marine snow mass and a large number of aggregates does not necessarily lead to a flat particle size distribution with many large particles. In the second half of the 5th year of the simulation the number of marine snow aggregates is increasing at 175°W 55°S, but the mean size of the aggregates is small, because only a small fraction of phytoplankton are diatoms, and hence the size distribution has a steep slope. The next maximum of the marine snow mass and aggregate number is much stronger and the size distribution is flatter with more large particles, which sink and reduce mass and number of aggregates. Because the size distribution exponent varies in space and time, average sinking speeds also vary with space an time. In large parts of the world oceans mean sinking velocity in the euphotic zone is lower than 5 m/d on average. But, depending on season and location also significantly higher values are found. In northern hemisphere winter high mean val-

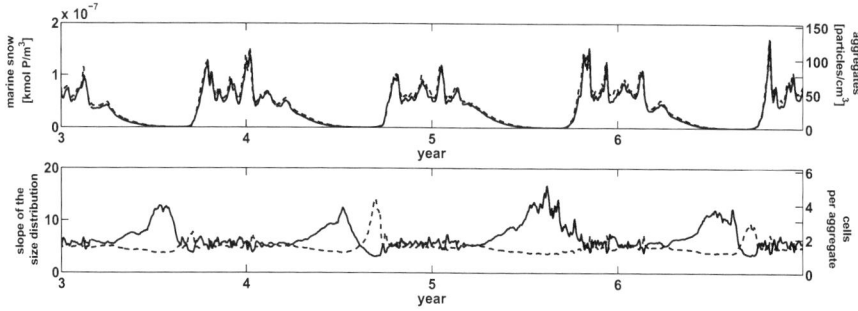

Fig. 2.8: Time series of mean marine snow concentration [kmolP/m³] (solid), number of aggregates [particles/cm³] (dashed), and the mean slope of the particle size distribution in the euphotic zone (solid), and number of aggregates (dashed) at 175°W 55°S.

ues of 15 m/d and more are found in the Southern Ocean. In northern summer the high abundance of organic material, especially diatoms due to the phytoplankton blooms leads to aggregation of large particles and high mean sinking velocity consequently. In the storm tracks of the southern hemisphere mainly low mean sinking velocities are found due a steep particle size distribution (not shown here), which is caused by sinking of large particles out of the euphotic zone.

Comparison and evaluation of biogenic tracers Besides differences arising from the implementations of the aggregation module and the satellite data assimilation the two experiments differ in some parameters affecting remineralisation of detritus, phytoplankton mortality below the euphotic zone and dissolution of calcium carbonate and opal in the total water column (Table 2.3). Open ocean regions can serve for an evaluation of the impact of aggregation versus constant sinking and the differing parameters on oceanic tracers relevant for DDT cycling. DDT in the colloidal phase is defined as the amount partitioning to the sum of phytoplankton, zooplankton and dissolved organic carbon. The sum of these phases (in the following referred to as colloidal organic matter, COM) integrated over the euphotic zone shows significant differences between the experiments in the northern hemispheric summer and winter (Figure 2.10). The main spatial structure is largely dictated by the availability of nutrients fuelling primary production. In both seasons COM concentrations are high off the west coast of Southern Africa and South America, and in the equatorial Pacific, due to wind-induced upwelling of cold, nutrient-rich water from the oceans interior to sun-lit surface waters. In summer in the northern hemisphere, north of 40 ° N, high insolation together with warmer temperatures trigger a phytoplankton bloom. Here sufficient nutrients are available, as they accumulated in the winter period when lacking insolation suppressed phytoplankton growth. Low values are found in the subtropical gyres, where Ekman downwelling and a stably stratified pycnocline limit reentrainment of nutrients into the sun-lit surface layer, and where even in winter the surface mixed layer does not significantly exceed the depth of the euphotic zone [Oschlies (2008)]. The main patterns are similar. The experiments, however, differ in the intensity of the concentrations: These are lower in the experiment with formation of marine snow aggregates almost everywhere and in all seasons. Low biomass can be explained by the higher phytoplankton mortality in the aggregation run, except for regions with high sinking velocities, in which a significant amount of biomass might have been removed from the euphotic zone, and hence low colloidal organic matter concentrations remain.

The spatial distributions of modelled phytoplankton concentration agree only moderately with the patterns of the satellite based observations (Figure 2.11). For the analysis both, Case 1 and Case 2 chlorophyll-a concentrations, are converted into their phytoplankton equivalents using a constant carbon to chlorophyll ratio, as it was done in the assimilation. Open oceans are compared with Case 1 product and coastal zones with Case 2 product. High phytoplankton values derived from Case 2 chlorophyll-a concentrations for coastal zones are strongly underestimated by the model experiments. The experiment with satellite data assimilation shows higher concentrations in comparison with the aggregation experiment also in coastal zones, as already discussed in section 2.4.3 (Figure 2.7). In open ocean regions the aggregation experiment is closer to the observations showing lower concentration values. In the equatorial upwelling regions both model results show higher phytoplankton concentrations than the satellite data.

The seasonal variability of phytoplankton is shown in Hovmöller diagrams (Figure 2.11) with the zonally averaged phytoplankton concentration in the first model layer plotted versus time. Both model prediction and satellite products suggest the highest seasonal variability in the higher latitudes. In the MPI-MCTM the phytoplankton maxima are very short, indicating a primary production occurring in relatively short pulses. The zonal mean maxima are highest in the SAT experiment. Export produc-

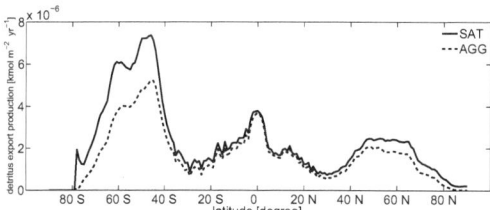

Fig. 2.12: Zonal mean annual detritus export production across 90 m [kmol m⁻2 yr⁻1].

tion describes the amount of particulate organic carbon that is transported from the surface across the 90 m level (depth of the euphotic zone in the model), and is of importance for estimating the vertical migration of DDT in the ocean. Highest export production occurs in the high latitudes between $40°$ and $60°$ north and south (Figure 2.12). The export production maxima are located in the zones of highest organic matter concentrations. The two experiments, AGG and SAT differ for the high latitude maxima, where the experiment with satellite data assimilation shows significantly higher export production than the aggregation experiment. The equatorial maximum is equally strong in both experiments.

Sinking and vertical export In the two experiments, AGG and SAT ocean dynamics and hence also vertical water movements are equal. Differences in vertical stratification of DDT can therefore be caused only by differing settling of DDT with organic matter at various ocean depths. Deposition from the atmosphere is the only source of DDT for the ocean. Thus, partitioning of DDT to organic phases in the euphotic zone determines its fate in the water column. In the satellite assimilation experiment, the mean fraction of DDT bound on COM (phytoplankton, zooplankton and dissolved organic carbon) in the euphotic zone is higher, caused by the higher abundance of colloidal organic matter (Figure 2.13). Largest differences occur in the Southern Ocean, the southern hemisphere storm track zone, and in the euqatorial upwelling regions. Those are at the same time the regions in which generally the highest partitioning of DDT to organic matter occurs. In the SAT experiment approximately 15- 25 % are bound to colloidal organic matter and 1.2-1.5% to particulate organic matter in the Southern ocean and the storm track zone. In the AGG experiment, on the other hand, the band of maximal fractions on colloidal organic matter in the Southern Ocean is much narrower and maxima reach only 15 %. The amount bound to particulate organic matter is highest west off the Peruian Coast, and in the equatorial upwelling region of the Pacific Ocean. In AGG approximately 0.7-1.4 % are bound to particulate organic matter, whereas in SAT $\approx$1.4-2.0 % of DDT are in the particulate organic phase.

Despite the large differences of DDT fractions in organic matter, spatial patterns of DDT below the euphotic zone are similar in both experiments. Locally, up to over 90 % of DDT is found below the euphotic zone. At the equator and in regions of coastal upwelling this fraction is less than 50% (Figure 2.14). Largest fractions of DDT below the eupotic zone correspond to regions of downwelling water due to Ekman pumping in the Atlantic and Pacific Ocean around 30 ° north and south. Minima of DDT below the euphotic zone also correspond to dynamical features, such as coastal and equatorial upwelling. This explaines the similarity of the spatial patterns of AGG and SAT.

To assess gravitational settling of DDT, the settling flux F_s, is diagnosed by the amount of DDT bound to detritus [kg(DDT)/kmol(P)] in the lowest euphotic zone layer (75 m - 90 m) and the detritus export production F_e [kmol(P)/m^2]

$$F_s = \frac{c_{DDT.POC}}{c_{POC}} \cdot F_e \qquad (2.4)$$

The individual effect of particle settling and sinking by vertical water movement is difficult to discern, since both depend on the DDT concentration. Settling with particles acts only to particle-sorbed DDT, in contrast to turbulent mixing and downwelling which affects also dissolved and colloidal sorbed DDT. In the coastal area off the Peruvian Coast, for expample, detritus export is effective and so is the amount of DDT bound to organic matter. But upwelling counteracts the gravitational settling. Gravitational settling is especially important in regions like the Northern Pacific Ocean, where the vertical water velocity is low, but significant gravitational settling occurs (Figure 2.14). In AGG where less mass is bound to particulate organic matter, and DDT settling is less effective than in SAT, approximately 40 % of DDT is found below the euphotic zone in contrast to SAT, where 50-60 % are found below the euphotic zone. Also on the global scale gravitational settling out of the euphotic zone is more effective in the SAT experiment than in the AGG experiment.

The amount of DDT reaching the sea floor is estimated from the amount of DDT degraded in the sediment. In the continental shelf zones the amount of DDT degraded in sediment is higher in the satellite data assimilation experiment than in the experiment with aggregation of marine snow (Figure 2.15). Export production and gravitational settling are much higher in the SAT experiment, due to higher abundances of organic carbon. Although detritus export production and, hence, gravitational settling out of the euphotic zone was shown to be much higher in SAT, the amount of DDT reaching the seafloor in the open ocean regions is higher in AGG. The differences are strong in the Southern Ocean and Northern Atlantic Ocean. An arbitrary transect through the Atlantic Ocean shows that in the aggregation experiment a higher amount of DDT is bound to particulate organic carbon below the euphotic zone (Figure 2.16) . Due to the higher sinking velocities (up to 70 m/d) in AGG, remineralisation is reduced and particulate organic carbon is efficiently transferred to deep ocean layers.

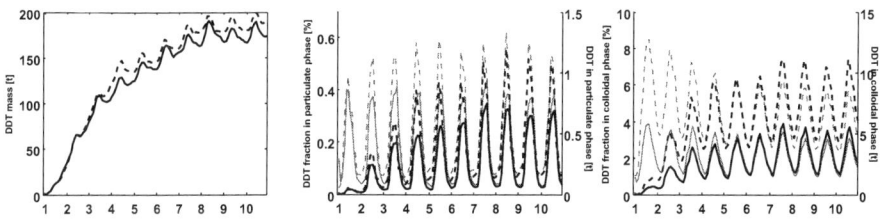

Fig. 2.17: DDT mass [t] and mass fractions [%] bound to colloidal and particulate phases integrated for all continental shelves. Grey lines show mass fractions, and black lines mass; dashed lines show results from the SAT, and solid lines results from the AGG experiment.

Continental shelf regions Approximately 7% of the total DDT mass is found in continental shelf zones (Figure 2.17) in both experiments. In general approximately 2-6 % of DDT mass on continental shelves is bound to the colloidal phase and approximately 0.6 % is bound to the particulate phase. This estimate is not based on the average fraction bound at individual locations but derived from total integrated DDT mass in the shelf zone bound to the particular phase. Fractions in individual shelf regions are higher (Figure 2.13). Continental shelf regions in the satellite assimilation experiment have been shown to contain more organic matter than corresponding regions in the aggregation experiment. The fractions of DDT bound to organic phases are larger in the SAT experiment consequently. In total approximately 4 % are found on average in colloidal phase in the SAT experiment and 2 %

in the AGG experiment. The fraction of DDT bound to particulate matter lead to a
larger amount of DDT in the water column between 50 m and 200 m (Figure 2.17),
and a larger amount of DDT transferred into the sediment.

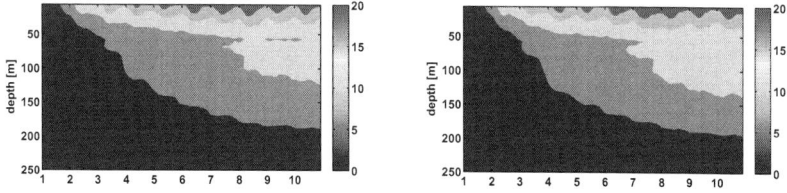

Fig. 2.18: Time series of vertical distribution of DDT mass [t] integrated for all continental shelves.
Left AGG experiment, right SAT experiment.

Global fate and distribution The global fate and distribution of DDT is characterised in terms of perstistence, long-range transport potential, and spatial and compartmental distributions. All relevant input parameters, e.g. location, time, and compartmental mass split upon emission (mode of entry) of the substance and degradation parameters, and the climate state are equal in both experiments. For that reason
the resulting differences between AGG and SAT reflect only differences in the spatial distribution of DDT caused by the differing representations of organic matter in
the ocean. Semeena (2005) and Guglielmo (2008) performed experiments using the
same substance scenario as presented here in different setups of MPI-MCTM. Un-
like in the current setup and the one used by Guglielmo (2008), the MPI-MCTM version used by Semeena (2005) only included a two-dimensional mixed layer ocean
neglecting transport and biogeochemistry. The comparison with results from these
two studies allows to assess the significance of differences between the experiments
AGG and SAT compared to impacts such as a different climatic state, spatial resolution, and representation of ocean dynamics. All results of the aforementioned
studies that are discussed here are summarised in Chapter 4 of Guglielmo (2008).

Principally the mass budget, compartmental and spatial distributions of both experiments, AGG and SAT, are very much alike (Figure 2.19, Figure 2.21). At the end
of the simulation most of DDT mass is found in soils (∼60%), followed by vegetation (∼23 %), ocean, (∼15%), and atmosphere (∼2%) (Table 2.4). Also in the
experiment by Guglielmo (2008) the compartment with the highest burden is soil.
Her results show a different distribution among the compartments, however, whith
52 % stored in soils, 27 % in ocean, 20 % on vegetation and 1% in the atmosphere.
Differences between the SAT and AGG experiments are small in comparison with
the differences of both experiments with the one conducted by Guglielmo (2008).
In the SAT experiment more mass is stored in the ocean and less in soil than in the
AGG experiment (Table 2.4). The higher fraction of DDT bound to organic matter
in the SAT experiment reduces the substance flux from the ocean into the atmo-

sphere and thus enhances the retention capacity of the ocean, i.e. persistence of DDT. The higher loss to the atmosphere is not compensated by the higher degradation of DDT in sediments in the AGG experiment, because volatilisation fluxes are much higher than sinking fluxes of DDT. Persistence is assessed by calculating the compartmental and overall residence times from the compartmental burdens and sinks in quasi-steady state. Soil and ocean don't establish a steady state within the simulated period of time. Therefore, residence time is calculated from mean burdens and annual sinks of the last four years of the simulation. The residence time calculated this way is smaller than the one in quasi-steady state, as both compartments are still accumulating mass. The residence times of DDT in the ocean is approxime-

Table 2.4: Compartmental distribution at the end of the simulation [%] and residence times [d] in quasi-steady state.

Compartment	Mass fraction [%]		τ [d]	
	SAT	AGG	SAT	AGG
Atmosphere	2.3	2.0	20.8	20.8
Soil	60.3	62.0	419.8	429.1
Vegetation	22.3	23.0	146.0	146.0
Ocean	15.1	13.0	449.0	405.2
Total	100.0	100.0	365.3	352.7

tely 43 days longer in the SAT experiment than in the AGG experiment. Guglielmo (2008) derived a residence time, in the ocean approximately 540 days, which is more than 100 days longer than the residence time estimated from current experiments. Generally two factors affect the residence time in the ocean, namely the spatial distribution determining volatilisation and the amount degraded in sediment. In Section 2.3 it was shown, that T42 does not resolve all continental shelves and water depth in resolved shelves is deeper than in GR15. Thus, in comparison with the current experiments, less mass would be degraded in sediments, which translates into a higher residence time. Furthermore the spatial distribution in the ocean differs from the current model results (Figure 2.20). In Guglielmo (2008) experiments a higher fraction of DDT is stored in high latitudes, where volatilisation is reduced due to colder surface temperatures. This factor again leads to a higher residence time in the ocean.

In the individual compartments quasi-steady state is achieved depending on emissions, degradation rates and spatial distribution of DDT. According to the seasonality of the parameters affecting degradation rates, e.g. temperature and oxidant abundance, the compartmental burdens in steady state follow a seasonal cycle. As the sources and consequently most of the DDT mass is located in the northern hemisphere, the cycle is defined by the climate of that hemisphere. Times needed to to achieve quasi staty state in the compartments are equal in the AGG and SAT experiment, as well as amplitude and phase of the burden time series. Vegetation reaches quasi-steady state within 2-4 years, and atmosphere already within 2 years. These times are in line with what was found by Semeena (2005) and Guglielmo (2008).

The long-range transport potential (LRTP) of DDT in the two experiments in the following is characterised using indicators introduced by Leip and Lammel (2004). Only indicators for potential LRTP are analysed. They describe the potential spatial mobility of a substance disregarding the fact that part of the substance is degraded in the environment. Of particular interest are the centre of gravity , the plume displacement (including zonal and meridional displacement) , the spatial spreading, and spatial scale of the distribution. The plume displacement (PD) is defined as the distance between the centre of gravity (COG) of the applications and the centre of gravity of the environmental burden at a particular point in time. Zonal and meridional displacements (ZD, MD) are defined as the distances between only the latitudes or the longitudes of these centres of gravity. The centre of gravity (COG) itself is given by the medians of the zonal and meridional cumulative distributions of the applications or burdens. Fifty per cent of the mass is found north- and southwards of its latitude, and in the hemispheres east and westwards of its longitude. With respect to the centre of gravity of the applications the COG of the environemtental burden is travelling approximately 12.5 ° to the East (ZD) and 8.5 ° to the North (MD) (Figure 2.21). Generally, the main migration characteristics of the COG of the total environment are determined by the COG in soil (Figure 2.22), the compartment storing the largest amount of DDT. In the first four years, however, the strong zonal variations are evoked by mass shifts in vegetation and ocean. In the ocean the centre of gravity undergoes strong zonal and meridional shifts in the first 3 years. Due to the fact, that DDT reaches the ocean only via deposition from the atmosphere, the ocean does not follow a smooth migration path like in soil and vegetation in the first years. From the forth year on migration of the COG within one year is predominantly in north-south direction, with a smaller migration to the east. In vegetation the centre of gravity migrates in north-west direction in the first half of the year and in southeast direction in the second half, when the northward migration is stronger than the southward migration, and the eastward motion stronger than the westward motion. In contrast to vegetation in soil the COG is moving to the northeast in the beginning and to the northwest in the second half of the year. Northward motion is stronger in the first half of the year. From the fourth year on north- and eastward migration gradiually reduce until the spatial distribitions reaches a quasi-steady state, indicated also by a stable, seasonally varying spatial spreading and spatial scale (Figure 2.23). In total the plume displacement is 1527 km in the satellite experiment and 1539 km in the aggregation experiment.

2.4.4 Summary

The global distribution of DDT has been simulated in two experiments that differ only in the representation of suspended organic matter in the ocean allowing to assess its impact on LRTP of DDT. Both experiments attempted to make results more realistic regarding the suspended matter representation. The first experiment

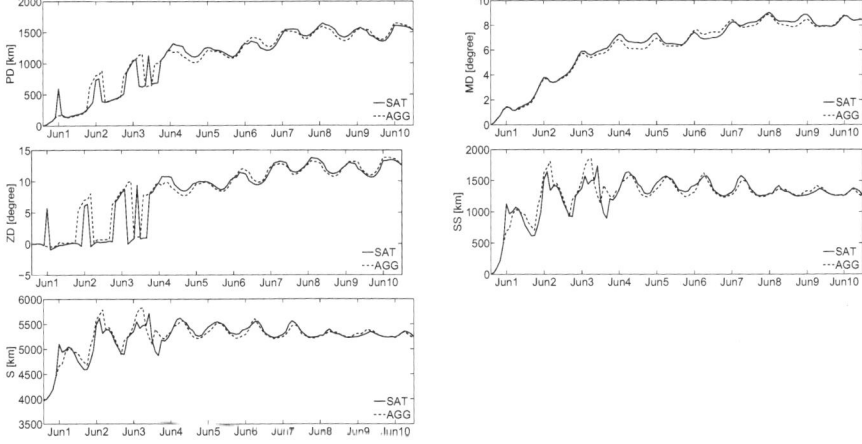

Fig. 2.23: Long range transport indicators: plume displacement (PD) [km], meridional displacement (MD) [degree], zonal displacement (ZD) [degree], spatial spreading (SS) [km], and spatial scale (S) [km]. A positive sign in MD and ZD indicates transport to the north or east, respectively.

simulated spatially and temporally variable sinking velocities of detritus based on a varying particle size distribution. In the second experiment HAMOCC was forced with satellite data on the oceanic margins imposing high coastal zone phytoplankton concentrations. Phytoplankton concentrations in the shelf regions were much higher than in the aggregation experiment, but due to the design of the experiment and the short simulation period the high gradient in the continental shelf areas was not fully established. In the SAT experiment the transfer of DDT to the sediment of continental shelves was more efficient than in the AGG experiment.

Differences in the representation of suspended organic matter lead to strongly differing organic matter abundances and amounts of DDT partitioning to it. The simulation of aggregation of marine snow lead to much more realistic phytoplankton concentrations in the surface ocean in comparison with satellite derived data. Similar to what was found by Guglielmo (2008) regionally up to 30 % of the DDT is bound to colloidal organic matter. The comparison of the experiments show compensating effects of organic matter on global DDT cycling. In the satellite experiment a higher fraction of DDT is bound on organic matter in the surface ocean thereby reducing volatilisation, but at the same time less mass is reaching the sea floor and degrading in sediment in open oceans, due to the lower detritus sinking velocity in the deep ocean. These effects do not fully cancel and induce differing compartmental distributions and residence times of DDT in soil and ocean. However, these differences among the experiments, are much smaller than differences in comparison with previous experiments with a less process resolved ocean compartment [Semeena (2005), Guglielmo (2008)]. Especially the comparison with Guglielmo (2008) underlines the higher importance of parameters, such as temperatures and circulation in atmosphere and ocean: These differ significantly from the current experiment, because of differing spatial resolution and restart conditions. Organic matter representations were shown to have litte impact on LRT of DDT. Differences in travelling distance, i.e. plume displacement, spatial spreading and spatial scale between the experiments are neglectibly small. No statistical evidence was supplied to show if the differences were significant or not. For this purpose either an ensemble simulation with slightly varying inital conditions or a longer simulation (assuming ergodicity) would be needed.

2.5 Sensitivity of volatitilisation of DDT from the ocean to climate parameters

2.5.1 Introduction

Volatilisation from the ocean surface plays an important role in the cycling of semivolatile organic compounds [Semeena and Lammel (2005)]. The air-sea exchange is thought to be controlled by wind speed, temperature, relative contaminant levels in air and water and the extent of association of POPs with particulate matter in the atmosphere and water column. Sea surface temperature affects the pollutants vapour pressure and solubility, consequently its ability to volatilise. The nonlinear relationship between sea surface temperature, wind speed and volatilisation rate suggests that the influence of these parameters on the volatilisation rate will vary for different temperature and wind speed regimes, hence in different climatic zones. Previous modelling studies (e.g. Wania and Mackay (1995)) quantified the significance of volatilisation and deposition based on zonal mean temperatures and wind speed. Here the impact of environmental conditions on variation of volatilisation rate in different parts of the global ocean is examined with a spatially resolving model. A simple analysis of the parameterisation of the volatilisation in the model shows, that for preset changes in temperature and wind speed the change in volatilisation rate is greater in areas with a high mean sea surface temperature. By analysing the impact of wind speed, sea surface temperature and pollutant concentration on the volatilisation rate in terms of a correlation analysis the prevalent wind speed and SST regimes will be isolated.

2.5.2 Experimental setup

In this study a ten year model simulation with the fully coupled ocean-atmosphere GCM [Roeckner et al (2003), Marsland et al (2003), Maier-Reimer et al (2005), Lammel et al (2001)] was performed. The insecticide DDT was chosen because it is one of the most studied of all synthetic organic chemicals and because of its physicochemical properties. It has a low water solubility, high lipid solubility and it is semivolatile. The lipid solubility allows the substance to partition to organic matter. The vapour pressure of DDT enables the compound to get re-volatilised from the oceans surface and undergo several deposition-volatilisation cycles [Semeena and Lammel (2005)]. The model ran with fixed agricultural applications taken from the year 1980 based on country level data from the FAO scaled with the crop density distribution [Semeena and Lammel (2003)]. The substance was applied to the soil (20%) and vegetation (80%) compartments of the model. The only source of the pollutant for the ocean was deposition from the atmosphere. The model volatilisation from the ocean surface is based on the two-film model using the fugacity approach with trans-fer coefficients taken from Mackay and Yeun (1983) quoted in Schwarzenbach et al (2003). The volatilisation is expressed in terms of SST (t), wind speed (u) and the concentration of the pollutant in the dissolved phase (c):

$$V = D(f_o - f_a). \tag{2.5}$$

Gaseous deposition from the atmosphere is calculated separately within the atmo-sphere model, hence $f_a = 0$. With $f_o = Hc$ the equation reduces to:

$$V = DHc = \frac{cH}{\frac{Rt}{f_1(u)} + \frac{H}{f_2(u)}} \tag{2.6}$$

where H denotes the Henry coefficient, $f_{1,2}$ are empirical functions of the wind speed [see Schwarzenbach et al (2003)].

2.5.3 Results and discussion

As the volatilisation flux strongly depends on the absolute contaminant mass, the volatilisation mass flux divided by the total amount of DDT in the first level of the ocean model is examined instead. This parameter is called volatilisation rate. It reflects the proportion of the mass abundant in the oceanic surface layer that was volatilised within one model time step. It depends upon how much of the DDT is dissolved in water and upon wind speed and sea surface temperature. The volatili-sation on the other hand would mainly mirror the deposition and emission pattern, because those are superposed onto the volatilisation defining patterns and dominat-ing because of the stationary application in the scenario.

The 10 year mean volatilisation rate (Figure 2.24) shows a pronounced latitudinal structure following the pattern of the mean sea surface temperature (Figure 2.28). Highest rates are located in the tropical ocean (or so-called warm pool) with a local minimum along the equatorial upwelling regions. The volatilisation rate decreases in direction to the extratropics reaching its minimum in the Arctic and Antarctic Oceans. Besides this meridional pattern a strong east-west difference up to a factor of two can be observed in the Atlantic and Pacific Oceans. The structures match the positions of warm and cold surface currents. High volatilisation rates are seen in regions where warm currents carry tropical water northward, for example the Kuroshio and the Gulf Stream. Corresponding low rates are found where cold waters move southward in the California Current and Canary Current.

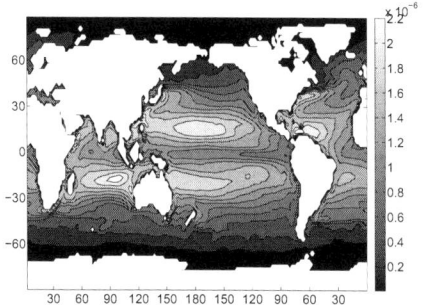

Fig. 2.24: Relative volatilisation mass flux 10 year mean [kg/(kg s)].

To estimate the impact of zonal averages versus zonally resolved SST and wind speed, volatilisation rates were diagnosed from formula 2.6 in two different ways. First the zonally resolved SST and wind speed predicted by the model was used to calculate a volatilisation rate, which was averaged zonally afterwards. Then, the volatilisation rate was estimated from beforehand zonally averaged SST and wind speed. The comparison of both results (Figure 2.25) shows, that the usage of zonally averaged SST and wind speed leads to an underestimation of the volatilisation rate and its variability in almost all latitudes, except for the high latitudes of the southern hemisphere. The underestimation is especially strong in the mid-latitudes of the northern hemisphere, where in particular the sea surface temperature shows significant zonal variation (caused by the warm and cold boundary currents).

The correlation coefficients between a 10 year monthly mean time series of volatilisation rates and SST, 1 0m wind speed and pollutant concentration are used to elucidate which of the parameters drives the volatilisation rate changes and causes the deviations from the long term mean. All of the parameters do not vary independently. Since both SST and wind speed influence the volatilisation rate in a nonlinear manner, it is not intuitive whether an increase in wind speed leads to an increase in volatilisation rate. A raise in wind speed that coincides with a decrease of the sea surface temperature can lead to a negative linear correlation coefficient between volatilisation rate and wind speed. For that reason the partial correlation coefficient is calculated in addition to the simple linear correlation coefficients. It explains the relation between a dependent and one or more independent parameters with reduced danger of spurious correlations due to the elimination of the influence of a third or fourth parameter, by holding it fixed. One important feature of the partial correlation coefficient is, that it is equal to the linear correlation coefficient if both variables

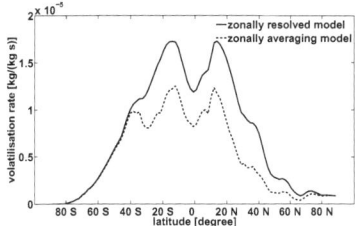 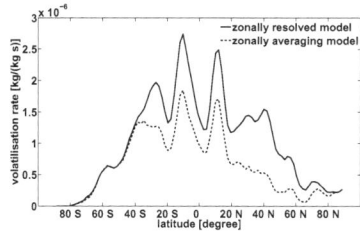

Fig. 2.25: Temporal mean (left) and standard deviation (right) of the zonal mean volatilisation rate over 10 years [kg/(kg s)]. Dashed lines show volatilisation rates derived from zonal mean SST and wind speed (denoted as zonally averaging model). Solid lines show volatilisation rates derived from zonally resolved SST and wind speed, which were zonally averaged afterwards (denoted as zonally resolved model).

are uncorrelated with the parameters held fixed in the calculation of partial correlation [Ellet and Ericson (1986)]. Thus, the partial correlation coefficient [2] between SST (t) and volatilisation rate (v) $R_{tv.uc}$ is equal to the linear R_{tv} if both SST and volatilisation rate are independent of wind speed (u) and pollutant concentration (c). This topic has never been studied with a global multicompartmental model.

In the following the coefficient of determination R^2 is used to find out which of the parameters explains most of the variance of the volatilisation rate. To illustrate the dominance of one or another coefficient of determination their values were interpreted as true colour values adding up to an RGB[3] image where the R^2_{uv} is associated with red, R^2_{cv} with green and R^2_{tv} with blue. Additionally the values were normalised to add up to one for reducing the colour spectrum to the shown triangle (Figure 2.26).

The RGB composite of the coefficients of determination of the individual linear correlation coefficients (Figure 2.26) shows that for the northern hemisphere high correlations of volatilisation rate and wind speed in the Atlantic Ocean can be found in the Gulf Stream and low values in the Labrador Sea and the adjacent Davis Strait. High correlations with the sea surface temperature are located near 45 °N close to the eastern coast of the American continent, in the Baltic Sea, North Sea and in

---

[2] Partial correlation coefficients are denoted by $R_{12.34}$, linear correlations by R_{12} substituting 1,2,3,4 with sea surface temperature (t), 10m wind speed (u), volatilisation rate (v) and pollutant concentration (c). Indices after the dot refer to variables held fixed in the calculation.

[3] RBG is an additive colour model in which each colour is described by how much of each of red green and blue is included in it. The colours are expressed as RGB triplets (r,g,b), with the individual components varying between 0 and 1. When all the components are zero the result is black; when all are one, the result is white. When all components are the same, the result is gray. When one of the components has the strongest intensity, the colour is near this primary colour (reddish, greenish, or bluish), and when two components have the same strongest intensity, then the colour is a secondary colour (cyan, magenta or yellow). A secondary colour is formed by the sum of two primary colours of equal intensity: cyan is green+blue, magenta is red+blue, and yellow is red+green. The RGB model is represented geometrically by a cube within the 0-1 range [Wolff and Yaeger (1993), Gonzalez and Woods (1993)].

the Northern Atlantic. In the Pacific Ocean the blue colour indicates that the coefficient of determination of the correlation between volatilisation rate and sea surface temperature, R^2_{vt}, is highest in most of the regions. Only in the subtropical part R^2_{vu} and R^2_{vc} are sufficiently higher than R^2_{vt} to cause red and green/yellow colours. The difference between the coefficients of determination of the partial correlation coefficients $R^2_{tv.uc} - R^2_{uv.tc}$ (Figure 2.27)a shows in the northern hemisphere a similar spatial pattern like the RGB composite of the R^2 of the linear correlation. The correlation between volatilisation rate and wind speed with fixed SST and pollutant concentration is larger than $R^2_{tv.uc}$ for the Gulf Stream and in those regions which show a dominance of the red colour in the RGB composite. Vice versa, $R^2_{tv.uc}$ is larger than $R^2_{uv.tc}$ in the Northern Atlantic, east of the American continent and in the major part of the Pacific Ocean. The differences between those two coefficients of determination are much smaller than the differences between the R^2s of the linear correlation. The reason for this is that both partial correlations are high (>0.8) in all of the northern hemisphere Atlantic and Pacific Ocean. On the east coast of the American continent and in the Hudson Bay the linear correlation coefficient between wind speed and volatilisation rate is very low and negative, but the partial correlation coefficient that excludes the influence of the sea surface temperature is 0.6-0.8. This is an indication for a strong suppression[4] of the expected positive correlation between wind speed and volatilisation rate by the influence of the sea surface temperature. Here, the standard deviation of the sea surface temperature (Figure 2.28) has its maximum, because of the strong annual cycle of the temperature of continental air masses formed over the Eurasian and the American continents migrating eastward. These air masses are warm in summer but cold and dry in winter. This causes an intense change of sensible and latent heat fluxes at the ocean surface close to the coast [Peixoto and Oort (1992)]. At the same time the mean sea surface temperature is very low. In the Gulf Stream the linear correlation between SST and volatilisation rate is close to zero, while the partial correlation coefficient is larger than 0.7. At the same time the correlation between wind speed and volatilisation rate exceeds 0.6 in the linear approach and 0.8 in the partial correlation analysis. Both cases imply a suppression of the linear correlation due to the influence of the variable which is held fixed in the partial correlation. Nonetheless, one can conclude from the RGB plot in Figure 2.26 in combination with Figure 2.27a, that it is the variation of the wind speed that determines the volatilisation rate changes.

In the northern hemisphere the coefficient of determination of the partial correlation between the pollutant concentration in the dissolved phase and the volatilisation rate excluding wind speed and SST, $R^2_{cv.ut}$, is very low in comparison to both coefficients that omit the pollutant concentration (Figure 2.27). Hence the apparently high correlation between pollutant concentration and volatilisation rate, shown as yellow to green colour in Figure 2.26 in some areas in the Pacific Ocean is not caused by a causal relation between them, but spurious. Both partial correlation coefficients are much lower there. The differences between values of $R^2_{tv.uc}$ and $R^2_{cv.ut}$ (Figure 2.27c)

---

[4] Relations between linear and partial correlation coefficients are described in Appendix A2

are very low in the Tropical Pacific, because also the coefficient of determination $R^2_{tv.uc}$ is lower than 0.3 in that area.

2.5.4 Summary

From the comparison of the linear and partial correlation coefficients it became obvious that the linkage between SST and wind speed has a strong impact on both the relation between volatilisation rate and wind speed, and volatilisation rate and SST. Nonetheless, the same spatial pattern of the explained volatilisation rate variance by either sea surface temperature or wind speed are found. These corroborate the hypothesis that the mean sea surface temperature represents the prevalent driving force for changes of the volatilisation rate. They, however, also indicate a dominating influence of the wind speed on changes of the volatilisation rate in regions of high mean SST, such as the Gulf Stream or the tropics, and a stronger control of the SST in colder areas. Obviously, a high variability of sea surface temperature like the one found east of the continents is able to amplify the effect of the sea surface temperature in comparison to the wind speed.

Basing models on zonal mean wind speed and sea surface temperature is averaging over regions where very different regimes of interaction of wind speed, SST and volatilisation rate prevail. In a long-term mean this leads to an underestimation of the volatilisation rate and its variabilty.

Since driving force of variations of the volatilisation rate is influenced by the predominant mean sea surface temperature changes of it will influence the evolution of the volatilisation rate and, hence, the distribution of the substance. The influence of the wind speed is expected to increase in a warming climate with higher sea surface temperatures, as it was shown that for high sea surface temperatures the variance is dominated by wind speed changes.

Chapter 3
Model application

3.1 Environmental fate and distribution of DDT: Model simulation with transient historical applications

3.1.1 Introduction

Dichlorodiphenyltrichlorethane (DDT) is a contact and stomach insecticide first used during the Second World War for control of lice and mosquitoes to combat typhus and malaria [Snedeker (2001)]. From the 1950s on, it was excessively used on a variety of agricultural crops worldwide. DDT is toxic to freshwater and marine microorganisms, fish, amphibians and birds [Ritter et al (1995)]. DDT is moderately to slightly toxic to mammals. DDT, mainly its metabolite DDE has been linked to reproduction disorders, like thinning of the eggshell of birds [Ritter et al (1995)]. Upon a classification by the World Health Organisation (WHO) DDT is categorised as moderately hazardous, based to its acute oral LD_{50} [1] for rats of 113-450 mg/kg [IPCS (1979)]. It mainly affects the central and peripheral nervous systems, and the liver. Acute effects in humans exposed to low to moderate levels may include nausea, diarrhoea, increased liver enzyme activity, irritation of the eyes, nose and/or throat. At higher doses, tremors and convulsions are possible [Ritter et al (1995)]. DDT is slowly biodegraded, persistent, and accumulates in fatty tissue of living organisms and along the food chain.

Growing concern about adverse environmental effects led to severe restrictions and bans in many developed countries since the early 1970s. As one of the so-called 'dirty dozen' DDT is nowadays banned in many nations, namely those who ratified the Stockholm Convention [2][UNEP (2001)], although it is still used in developing

---

[1] LD_{50} is an index of toxicity defined by the amount of substance killing 50% of a test population (LD = Lethal Dose).

[2] The Stockholm Convention is a global treaty to protect human health and the environment from persistent organic pollutants (POPs). It was adopted on 22 May 2001 and entered into force on 17 May 2004. Official website:.http://chm.pops.int/.

countries due to its cost effectiveness and broad-spectrum activity. Under the recommendation of the WHO the Stockholm Convention permits usage of DDT for malaria vector control, such as for indoor residual spraying, mainly because of the absence of equally effective and efficient alternatives to it and a simplified resistance management, because resistance development to DDT is no longer influenced by other uses [WHO (2006)].

Despite the restrictions to usage and production there still remain extensive amounts of DDT and its metabolites in the environment. Traces of DDT were detected all around the world, including regions where there are no direct sources, such as open ocean waters [Tanabe and Tatsukawa (1983),Iwata et al (1993)], the Arctic [Patton et al (1989)], and high mountain areas [Villa et al (2003)].

Within the last 30 years numerous studies have been conducted to understand transport pathways of DDT and other multicompartmental substances in the environment. In particular, models of varying complexity (temporal, spatial and process resolution) [a.o. Cramer (1973), Wania and Mackay (1995), Lammel et al (2001), Klöpffer and Schmidt (2003), Semeena (2005), Schenker et al (2008a)] were used to study processes affecting long-range transport, distribution, and fate. DDT is semi-volatile ($10^{-6} < p_{sat} < 10^2$ Pa), facilitating enhanced atmospheric transport by multihopping [Wania and Mackay (1993), Semeena and Lammel (2005), and parameters such as regional climate, location of the sources, and ocean circulation are of importance for its migration [Semeena and Lammel (2003), Leip and Lammel (2004) Guglielmo (2008)]. The importance of these spatially varying parameters implies limitations of multimedia mass balance box models (MBMs) in assessing long-range transport and fate of DDT due to no or low spatial resolution. Models with high spatial resolution, like MPI-MCTM, however, were not used so far to reconstruct the distribution of DDT using realistic applications in a long-term simulation.

This study represents a simulation of DDT and its metabolite DDE of 40 years with reported agricultural applications. Model results for the different environmental compartments are evaluated by a comparison to observations

DDT/DDE physical and chemical properties Technical grade DDT is a mixture of the isomers p,p'-DDT, o,p'-DDT and o,o'-DDT, where p,p'-DDT makes up more than 80% of the mixture. DDT is highly insoluble in water and is soluble in most organic solvents. It has a saturation vapour pressure of $2.5 \cdot 10^{-5}$ Pa at 293 K, hence it is semivolatile. DDE (1,1-dichloro-2,2-bis(4-chlorophenyl)ethylene) is formed from DDT through photochemical reactions [Maugh (1973)] and under aerobic conditions through dehydrochlorination in bacteria and animals [Aisalbie et al (1997)]. A list of the relevant substance properties is given in Table 3.1.

DDT applications During the Second World War DDT was used to combat typhus and malaria, which was achieved with relatively small amounts [Turusov et al (2002)]. After the war DDT was used for the control of pest in agriculture and forests, and in health programs. Accumulated total use of DDT between 1950 and 1993 was approximately 2 600 000 tonnes [Voldner and Li (1995)]. Among the

Table 3.1: Physico-chemical properties of DDT and DDE.

Physico-chemical property	DDT	DDE
Degradation rate in soil [1/s]	4.05e-9[a]	0.00
Degradation rate in ocean [1/s]	0.00	0.00
Saturation vapour pressure [Pa]	2.5e-5 (293 K)[f]	8.7e-4 (303 K)
Enthalpy of vapourisation [kJ/mol]	118[b]	92.818
Water solubility [mg/L]	3.4e-3 (298 K)[c]	0.4e-1 (293 K)
Enthalpy of solution [kJ/mol]	27[b]	160
Henry coefficient [Pa m^3/mol]	0.35185 (298K)[d]	4.2 (298K)
Octanol-air partitioning coefficient (logKoa)	10.09	9.69
Octanol-water partitioning coefficient (logKow)	6.19	5.89
OH gas-phase rate constant [cm^{3}molec^{-1}s^{-1}]	1.0e-13[b]	2.0e-11

[a] Hornsby et al (1996)
[b] estimated
[c] Biggar et al (1967)
[d] calculated from vapour pressure and water solubility [3.4e-5mg/L (298K)]
[e] O'Brien (1975)

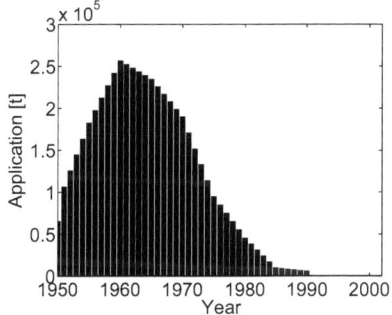

Fig. 3.1: Annual global DDT applications [t] [Semeena and Lammel (2003)].

countries with the highest usage are the United States, the Soviet Union and China [Semeena (2005)].

As there are no direct uses of DDE, it is found in the environment only as a result of contamination and breakdown of DDT.

The model was forced with agricultural application data of the insecticide DDT compiled by Semeena and Lammel (2003). Statistical data of DDT consumption reported by member of the UN states to Food and Agriculture Organisation (FAO) were combined with other published data (details in Semeena and Lammel (2003)). The emission inventory assumed 100 % of p,p'-DDT. After scaling the DDT consumption with crop land distribution, the data were extrapolated to the model grid. The result was a data set with spatially and temporally varying applications (accumulated application and temporal evolution shown in Figure 3.1). No seasonal or diurnal variation of the applications is considered.

DDT and DDE observations Iwata et al (1993) determined concentratons of DDT in in the air and surface water from various oceans in 1989-1990. The routes of the ship cruises among others covered the Chukchi Sea, Bering Sea, Gulf of Alaska, North Pacific, East China Sea, South China Sea, North Atlantic ocean, Bay of Bengal, and the Southern ocean.

Patton et al (1989) collected air samples on a floating ice island in the Beaufort Sea, located about 50 km off Ellesmere Island at about 81 ° N, 100 ° W. The samples were collected in August-September 1986 and in June 1987. Organochlorine concentrations were measured, including p,p'-DDE and p,p'-DDT.

Vertical profiles and surface concentrations of DDT dissolved in the ocean were measured by Tanabe and Tatsukawa (1983). Water samples for depths down to 5000 m were taken on three cruises of the Ocean Research Institute, University of Tokyo and on a cruise by the University of Fisheries, Tokyo. The cruises were carried out between 1976 and 1981 in the Western Pacific, Eastern Indian and Antarctic oceans.

Schenker et al (2008a) evaluated the CliMoChem model soil concentrations against observed concentrations from various sediment and soil studies. Their compilation of observed data was used to evaluate MPI-MCTM soil results.

Levels of DDT in pine needles were used for the evaluation of atmospheric DDT concentrations. In 1989 Hellström et al (2004) measured DDT concentrations in pine needles collected in various regions from Southern Germany to Northern Scandinavia. Jensen et al (1992) collected pine needles across Europe in 1986.

For an evaluation of the temporal trend of DDT in the Arctic ocean a comparison to selected trends of DDT in Arctic biotia was conducted. In particular trends of DDT in polar bears, ringed seal, and seabirds published in the AMAP assessment report [Arctic Monitoring and Assessment Programme (AMAP) (2004)] were analysed. The biopsy samples from adult female polar bears origin from the Churchill area of western Hudson Bay (Canada) and were collected in 1968, 1984, 1989, and nearly every year throughout the 1990s. Temporal trends of DDT in ringed seals were derived from female seals from the Canadian Arctic sampled at the communities of Ausuittuq (Grise Fiord), Ikpiarjuk (Arctic Bay), and Holman. Trends of DDT in seabirds eggs origin from the eastern Canadian Arctic at Prince Leopold Island. These data encompass three species of seabirds (black-legged kittiwake, northern fulmar, thick-billed murre) and cover the time period from 1975 to 1998.

3.1.2 Experimental setup

The three dimensional multicompartmental chemistry transport model MPI-MCTM [Lammel et al (2001), Semeena and Lammel (2003), Guglielmo (2008)] was run for 40 years with a resolution of T21L19 in the atmosphere and GR30L40 in the ocean. The ocean biogeochemistry started from spin-up fields and the atmosphere from an initial run, necessitating a physical spin-up of 2 years prior to the actual simulation. Within a run of 40 years the model produces its own climate based on pre-industrial CO_2 concentrations. At the beginning of the run the model environment was not contaminated by DDT. Transient DDT applications for the years 1950-1990 were applied strictly to soil (20 %) and vegetation (80%) simulating only agricultural use. The application is temporally homogeneous throughout the year. The model sinks for DDT are reactions with radicals (OH) in the atmosphere, degradation in soil and vegetation and a constant loss rate in the lowermost wet layer of the ocean to mimic degradation in the sediment. All DDT degraded in soil is assumed to metabolise into DDE. This is the only source considered for DDE. For DDE only atmospheric reaction with OH radicals and dechlorination to DDMU [1-chloro-2,2-bis(p-chlorophenyl)ethene] in anaerobic conditions in the ocean provide environmental sinks.

To understand the impact of individual processes on the compartmental distribution of DDT, model runs with a non-steady-state, zero-dimensional, multimedia mass balance box model (MPI-MBM) [Lammel (2004)] were conducted in addition to MPI-MCTM experiments. Parameterisations of intra- and intercompartmental mass exchange and conversion process in MPI-MBM are similar to those in MPI-MCTM. A detailed description of differences and a comparison of both models can be found in Lammel et al (2007). The DDT emissions were the global mean temporally varying DDT applications for the years 1950 to 1990. A repeating annual cycle around constant mean temperatures was simulated. Surface and air temperatures differ by 14 K constantly.

3.1.3 Results and discussion

Environmental distribution After 40 years of continuous application of DDT to vegetation (80%) and soil (20%), 73% of the total mass present in the environment in December 1990 are stored in soil, 24 % in the ocean, 2 % in vegetation, and less than one percent in the atmosphere (Table 3.2).The high storage in soil is caused by its strong absorptive capacity of organochlorine compounds, which is related to its organic matter content. The only source of DDT and DDE in the ocean is deposition

Table 3.2: Compartmental distribution [%] and residence time [a] at the end of the MCTM simulation.

Compartment	Mass fraction [%]		τ [a]	
	DDT	DDE	DDT	DDE
Atmosphere	0.04	0.14	0.02	0.01
Soil	73.55	94.69	14.93	3.58
Vegetation	2.22	0.92	1.20	0.48
Ocean	24.19	4.25	23.70	2.36
Total	100.00	100.00	16.70	3.49

from the atmosphere. Once deposited, they instantaneously establish equilibrium with the abundant organic phases, and are subject to gravitational settling bound to sinking particulate matter (detritus). Vertical diffusion and advection (sinking, deep water formation) also re-distribute them within the water column. Once the pollutant reaches below the mixed layer depths only continental upwelling, and entrainment of water into the mixed layer during storm events can bring it back to surface waters, hence a significant part of sinking DDT and DDE is irreversibly removed from participating in air-sea exchange. In contrast to the DDT compartmental distribution the largest fractions of DDE by December 1990 are stored in soil (95 %) and in the ocean (4 %). DDE is not directly emitted, but its only source in the simulation is degradation of DDT in soils. Both substances are similar lipophilic, which explains a high retention capacity of soils. Only part of the DDE mass volatilises and reaches vegetation and ocean compartments subsequent to deposition from the atmosphere.

The most noticeable feature in the environmental distribution at the end of the 40 year simulation is the interhemispherical gradient, prominent for DDT and DDE (Figure 3.2). It shows much higher levels in the northern hemisphere than in the southern hemisphere. The increase of the contamination from low to high latitudes reflects the distribution of the sources. Within the 40 years from 1950-1990 largest amounts of DDT have been applied to North America and central European agricultural land, followed by East and South Asia. The most highly contaminated sites are North America, European and Russian soils, and the north of India. In the ocean largest DDT contaminations are found east of the North American continent in the Gulf Stream region, and the Arctic ocean. Lowest DDT burden is located in the Indian ocean and the southern Pacific ocean. In the Indian ocean local maxima are related to the vicinity of the source. The high abundance of DDT the Southern ocean, however suggests strong interzonal transport in direction of the poles. It has been shown [Leip and Lammel (2004), Semeena and Lammel (2005), Guglielmo (2008)], that DDT is transported in several deposition-volatilisation cycles depending on regional climate and environmental features (so-called grasshopping) to remote regions. The atmospheric transport is the dominant route for DDT to the Antarctic ocean. Transport with the North Atlantic current system has been shown to be of importance for migration to the Arctic ocean [Guglielmo (2008)].

The American and European soils, highest contaminated sites at the end of the simulation, are in 1990 already less contaminated than before DDT usage was restricted in the early 1970s. They reached their contamination maxima in the 60s and early 70s (Figure 3.4), about 5 to 10 years after the local application maximum (Figure 3.3). High contaminations in Asia can be linked to highest DDT applications in the beginning of the 1980's.

The globally integrated compartmental DDT burdens shows all compartments already passed their maximum contaminations. Integrated compartmental mass is following the temporal pattern of the applications (Figure 3.5). The applications reached a maximum in 1960 and decreased since then. The peak in the atmospheric DDT burden occurred in 1961 (1 year later), in comparison to vegetation in 1966 (6 years later), soil in 1973 (13 years later), and ocean 1977 (17 years later). The

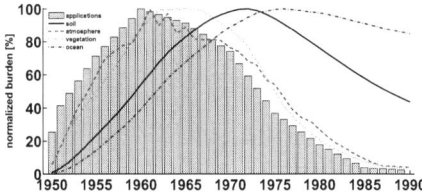 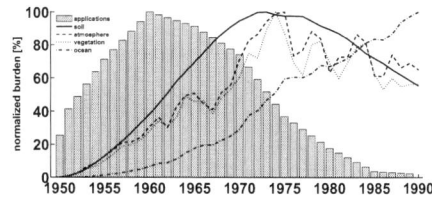

Fig. 3.5: Global compartmental burdens of DDT (left) and DDE (right) normalised to their maximum [%] for atmosphere, vegetation, soil and ocean, grey bars normalized global DDT application.

delay between the application maximum and the compartmental maxima reflects the mean residence time of DDT in the compartments (Table 3.2). The estimated residence times of DDT in the ocean is 18 years, in soils 14 years, in vegetation 1 year and in atmosphere around 7 days. The resulting overall residence in the environment time is 16.7 years (Table 3.2).

The vertical stratification of global DDT mass, integrated from the sea bottom upwards, shows that the global ocean below 100 m is constantly accumulating DDT (Figure 3.7). It is evident that only DDT mass in the uppermost levels is reaching maximum in the late 1977s and subsequently decreases. This indicates that only the upper oceanic layers work as a secondary DDT source to the environment after reaching the maximum in the late 1970s.

DDT shows a different behaviour in individual oceanic regions (Figure 3.4). Although the global DDT mass stored in the ocean is decreasing at the end of the simulation, the Tropical Atlantic and Pacific oceans in vicinity of the equator did not reach a maximum until 1990. Also parts of the Arctic ocean are continuously accumulating DDT mass, e.g. the Beaufort Sea. Other parts of the Arctic ocean reached a maximum and reside in a quasi-steady state, and some, like the Barents Sea have been losing mass since the beginning of the 1980s. The year of the maximal DDT burden in individual oceanic regions is defined by its distance from the source and the residence time of DDT within the region. The residence time of DDT in a region is defined by the local circulation pattern advecting DDT from other areas, the environmental conditions in surface waters affecting DDT air-sea exchange with the atmosphere, and the biological activity and vertical water movements eventually causing efficient removal of DDT from surface waters. DDT has the highest residence time in the Arctic ocean (Figure 3.6), in parts of which it is accumulating until the end of the simulation (Figure 3.4). DDT enters the Arctic ocean via atmospheric deposition and transport with the North Atlantic Circulation. On the other hand the Arctic ocean is losing DDT mass through advection into adjacent seas (e.g. the Baffin Bay and the Norwegian sea) and through volatilisation. Due to the low mean sea surface temperature volatilisation is reduced, and the Beaufort Gyre limits transport of DDT out of the Canadian Arctic. In contrast to DDT, DDE is constantly accumulating in the ocean, not only integrated over the world and in all levels, but also in all individual oceanic regions (evident in the Figures 3.7 and 3.5).

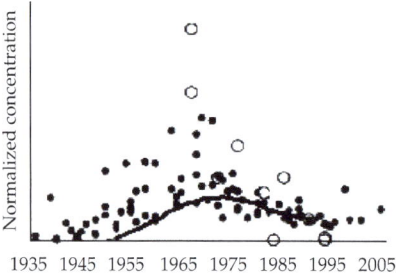

Fig. 3.7: DDT concentration in soil normalised to 1990 values.Solid line MPI-MCTM 40°N - 60°N, dots and circles observations from Dimond and Owen (1996), Meijer et al (2001), and others summarised in Schenker et al (2008a).

Comparison with observations Soil and vegetation are only represented as single layer (topsoil) surfaces in the MPI-MCTM, hence their contamination is expressed as a mass per surface area. Soil burdens were converted into concentrations by dividing them by soil dry bulk density and a fixed soil depth of 10 cm. The average DDT concentration in soil between 40°N and 60°N was compared to measured soil and sediment concentrations from Northern North America and Great Britain [Dimond and Owen (1996), Meijer et al (2001), and others compiled by Schenker et al (2008a)]. For intercomparison reasons only relative soil concentrations are compared to observational data. Each set of observations was normalised to its 1990 value.

Observations for DDT levels in US and European soils show that 1965 values of concentrations are about 2 to 6 greater than those in 1990 (Figure 3.7). Model data from MPI-MCTM show a peak in the concentrations around 1972 which are 2 times higher than 1990 values. Modeled and observed concentrations are decreasing from then on. For the second half of the simulation the model results are well within the observed range of relative concentrations, whereas for the first half model results are at the lower boundary of the range spanned by the observations.

Atmospheric concentrations in the model were compared to observations made over various ocean surfaces [Iwata et al (1993)] and on a floating ice island in the Canadian Arctic [Patton et al (1989)]. For the comparison with Iwata et al (1993) spatial means of the relevant ocean regions were derived and compared with the reported data. For the comparison with Patton et al (1989) the corresponding model grid box was identified.

Both observations and model results show a decrease in concentration from lower to higher latitude. For DDT, modeled concentrations, ranging between 6 and 700 pg/m^3, are higher than observed ones (Table 3.3), except in the Caribbean Sea and the South China Sea. Observed concentrations range between 0.9 and 230 pg/m^3. In the Bay of Bengal model results show a mean concentration of 712.5 pg/m^3, whereas the observed mean value is 250 pg/m^3. The model prediction lies within the range of the observations ranging from 42-1000 pg/m^3 [Iwata et al (1993)]. A better agreement is found in lower latitudes (e.g. the Gulf of Mexico) where model exceeds observations less than two times. Best agreements between model and observations are found in the Caribbean sea, the Gulf of Mexico, the Southern ocean and the South China Sea. In high latitudes of the northern hemisphere modeled concentrations are up to 100 times larger than observed values.

For DDE modelled concentrations range between 0.04 and 65 pg/m^3. Predicted concentrations are up to 100 times higher than observed concentrations in air over northern and Arctic seawaters (including the Gulf of Alaska, the Bering Sea, the Chukchi Sea, the North Atlantic, and the North Pacific oceans). In other evaluated oceanic regions, i.e. the Caribbean Sea, the South China Sea, the Gulf of Mexico, the Bay of Bengal, and the Southern ocean predicted concentrations are lower than observed, indicating that the discrepensies are not only caused by a general overestimation of the DDE source. In the East China Sea, however, the model overestimates the DDE contamination. Observations of DDT concentrations in pine needles [Jensen et al

Table 3.3: Atmospheric DDT concentrations c_{atm} in [pg/m^3], observations compared to model results. In the upper part observations are taken from Iwata et al (1993), in the lower part observations are from Patton et al (1989) who collected data at 81 °N 100 °W.

Location	Latitude	Concentration [pg/m3] obs. DDT	obs. DDE	model DDT	model DDE	Fraction [%] obs. $\frac{DDT}{DDT+DDE}$	model $\frac{DDT}{DDT+DDE}$
Gulf of Alaska	52.6 - 58.1 ° N	3.9	0.8	30.0	32.9	83	48
Bering Sea	52.6 - 69.2 ° N	3.6	0.8	37.6	44.3	70	46
Chukchi Sea	69.2 - 74.7° N	5.8	0.5	37.5	64.9	78	37
Gulf of Mexico	24.9 - 30.4 ° N	48.0	9.1	58.8	2.8	46	96
East China Sea	30.4 - 35.9 ° N	19.0	3.7	187.3	55.0	36	77
Caribbean Sea	13.8 - 19.4 ° N	13.0	6.4	9.1	0.4	35	96
North Atlantic	30.4 - 69.2 ° N	8.7	3.4	34.5	22.3	45	61
North Pacific	19.4 - 41.5 ° N	12.0	2.0	34.2	12.0	40	74
South China Sea	13.8 - 24.9 ° N	54.0	17.0	41.3	6.0	37	88
Bay of Bengal	19.4 - 24.9 ° N	250.0	19.0	712.5	0.4	56	>99
Southern Ocean	58.1 - 63.7° S	2.4	0.3	6.3	<0.01	50	99
Arctic (08-09/1986)	81.0 ° N	0.9	0.1	69.0	92.0	90	43
Arctic (06/1987)	81.0 ° N	2.3	2.9	58.7	4.7	44	93

(1992), Hellström et al (2004)] were used as a surrogate for atmospheric concentrations. They serve to evaluate the latitudinal gradients of the model results. The biomass DDT concentrations are compared to mixing ratios of DDT in the lowest atmospheric model layer. Ratios of observational data were calculated for pine needles from the same growth year.

Concentrations in Central and Southern Europe are higher than concentrations in Scandinavia for both model results and observations (Table 3.4). The ratio between Southern Germany and Northern Norway is 4.5 in the observations and 1.2 in the model. The ratio was derived from the pine needle data from Hallwagen and Zusamzell in Germany compared to data from Stronglandet, Norway [Hellström et al (2004)].

The ratio between Southern France and Central Sweden in pine needles [Jensen et al (1992)] is only slightly higher than the ratio of atmospheric concentrations for these regions (1.6 versus 1.4) (Table 3.4). The mean concentration from Toulouse, Bordeaux, Tarbes and Bayonne was compared to averaged data from various locations in Central Sweden [3] [Jensen et al (1992)].

The comparison of oceanic DDT concentrations was performed using two kinds of observational data. Surface concentrations for the years 1989 and 1990 reported as

Table 3.4: Latitudinal gradients of DDT in pine needles [Jensen et al (1992), Hellström et al (2004)] compared to ratios of atmospheric concentrations in the model.

	Observation	Model
Southern France / Central Sweden		
1986	1.6	1.4
Southern Germany / Northern Norway		
1989	4.5	1.5

area means of several ocean regions [Iwata et al (1993)] were used to evaluate the global spatial distribution of DDT. Vertical profiles from individual locations, measured in 1976 [Tanabe and Tatsukawa (1983)], were used to evaluate the modeled distribution of DDT in the water column.

For the comparison area means of the relevant oceanic regions (latitudes in Table 3.5) were compared to the mean reported concentrations. Surface water concentration reported by Iwata et al (1993) ranged from 0.3 to 10 pg/L. The mean area averages of the modelled concentrations ranged from 5 to 800 pg/L. Observed concentrations decrease from low to high latitudes, whereas modeled concentration strongly increase. In the Chuckchi sea deviations between model and observations are highest. For the selected regions measured concentrations are lowest (0.3 pg/L) in the Chuckchi sea, but modeled concentrations are highest (800 pg/L) in that sea. Generally deviations between model and observations are very high in the high latitudes of both , the northern and southern hemisphere.

For DDE predicted surface ocean concentrations are overestimated in the Gulf of Alaska, the Bering Sea, the Chukchi Sea, and the North Atlantic. In all of these regions also the atmospheric concentration was overestimated. This indicates that the strong overestimation in these regions is caused by enhanced deposition from the atmosphere, and amplified by low volatilisation due to low mean surface temperatures. Also in the East China Sea and in the Gulf of Mexico modelled surface

---

[3] Sampling locations are close to to Sundsvall, Ange, Ytterhogdal, Sveg, Tynset, Trysil, Gavle, Dala-Jarna, Malung, Salen, Avesta, Norrtalje, Hedemora and Sorentorp [Jensen et al (1992)]

Table 3.5: DDT concentration in the surface ocean in [pg/L], model results in comparison with observations from Iwata et al (1993).

Location	Latitude	Concentration [pg/m3] obs. DDT	obs. DDE	model DDT	model DDE	Fraction [%] obs. $\frac{DDT}{DDT+DDE}$	model $\frac{DDT}{DDT+DDE}$
Gulf of Alaska	52.6 - 58.1 ° N	1.2	0.2	290.0	19.2	75	94
Bering Sea	52.6 - 69.2 ° N	1.0	0.9	460.0	23.5	53	95
Chukchi Sea	69.2 - 74.7 ° N	0.3	0.2	800.0	324.0	60	71
Gulf of Mexico	24.9 - 30.4 ° N	2.2	0.3	17.0	41.0	88	30
East China Sea	30.4 - 35.9 ° N	16.0	3.0	71.0	11.0	84	88
Caribbean Sea	13.8 - 19.4 ° N	3.9	0.5	7.7	0.2	89	97
North Atlantic	30.4 - 69.2 ° N	0.8	0.5	135.0	10.0	62	94
North Pacific	19.4 - 41.5 ° N	1.2	0.5	5.1	0.4	71	93
South China Sea	13.8 - 24.9 ° N	6.9	1.0	8.5	1.5	87	85
Bay of Bengal	19.4 - 24.9 ° N	10.0	1.5	41.6	1.5	87	97
Southern Ocean	58.1 - 63.7 ° S	1.0	0.5	143.7	<0.1	67	>99

water concentrations are much higher than observed. In contrast to the overestimation in the East China Sea, the overestimation of concentrations in the Gulf of Mexico, can't be explained by overestimated atmospheric DDE concentrations, as these were shown to be underestimated. In the South China Sea and in the Bay of Bengal predicted surface water concentrations are in good agreement with observations despite discrepancies in air concentrations over these waters.

Vertical DDT profiles were compared with observations (see figure 3.9) at four locations in the Pacific, Indian and Antarctic oceans [Tanabe and Tatsukawa (1983)]. Measurements in the Pacific were conducted at two locations west of Japan in July (A) and August (B) 1976. At location A probes were taken at 0 m, 50 m, 200 m, 500 m, 1000 m, and 1500 m. Modelled and observed concentrations show a similar profile decreasing from the surface down to 200 m. Model results decrease from 1.5 ng/L to approximately 0.5 ng/L, whereas observations range from 0.75 ng/L to 0.25 ng/L. From 500 m to 700 m model results first increase and than decrease moderately down to 1000 m after reaching a maximum. The observations fail to capture this profile, because of resolution, i.e. no samples were taken between 500 m and 1000 m. Below 1000 m model results show a rapid decrease to 0 ng/L. Observations are increasing from 1000 m (0.58 ng/L) to 1500 m (0.67 ng/L). At location B the modelled profile shows a similar pattern like at location A. Both, model results and observations decrease from 1 ng/L (observations), 1.5 ng/L (model data) to approximately 0.4 ng/L at 100 m. As for location A the profiles cannot be compared directly because observations capture 7 depths between 0 and 2500 m. Due to the limited model resolution and consequential deficiencies in the model bathymetry (discussed for GR15 in Chapter 2), the model ocean is shallower than the deepest depth of 2500 m in the observations at location B. At location E in the Indian ocean southwest of Indonesia, model results and observations show a different profile. At the surface observational data are higher than model results, decreasing down to 200

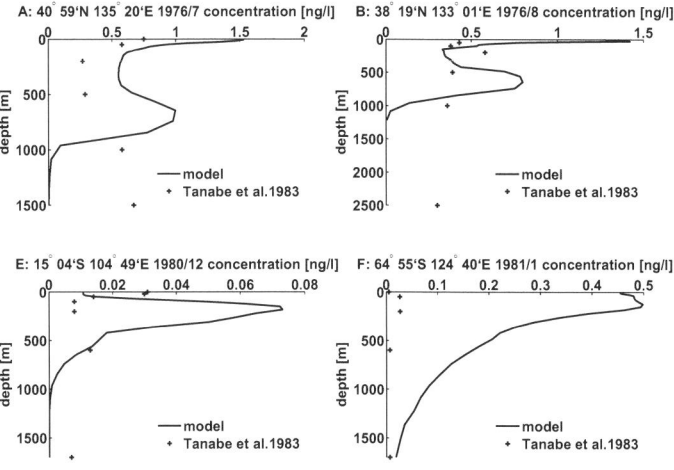

Fig. 3.9: Vertical profiles of DDT concentration [ng/L] in the Pacific ocean close to Japan (A),(B), in the Indian ocean (E), and in the Antarctic ocean (F). Model results in comparison with observations from Tanabe and Tatsukawa (1983).

m, and showing a slight increase at 600 m. The model results increase down to about 200 m and decrease afterwards. In the Antarctic ocean the model results exceed the observations for all depths. Both profiles show a subsurface increase down to approximately 100 m followed by a constant decrease. Maximal model results are 0.5 ng/L, whereas maximum observed concentrations are 0.027 ng/L.

Temporal trends of ΣDDTs (DDT+DDE) in Arctic biota were used to evaluate the predicted trend of ΣDDTs in seawater. For that purpose a linear regression of the ΣDDTs trend in the 1980s was calculated and used to derive the half life of ΣDDTs. The marine organisms analysed include polar bears, ringed seals, and seabirds. The half lives of ΣDDTs in the organism were compared to half lives derived from mean surface layer concentrations in ocean basins adjacent to the location where the sample was collected. All sampled of Arctic biota, and also the predicted surface layer concentrations revealed a significant decrease of ΣDDTs in the 1980. The apparent half life of ΣDDTs in female polar bear was 12 years (Figure 3.6). The trends observed in adult female polar bear are likely to represent the changes of ΣDDTs in the polar bear food chain. The Arctic Monitoring and Assessment Programme (AMAP) (2004) states that the strong decrease of ΣDDTs throughout 1968 to 1990s could be related to local conditions around Hudson Bay. An extensive biting-insect control was conducted in communities and military bases in the Hudson Bay area in 1950s and 1960s, which may have contributed to a significant load to the Hudson Bay. This source is not represented in the emission inventory used for the model simulations. The half life derived from surface water concentrations in the Hudson Bay shows a more rapid decline of Σ DDT indicated by a half life of 9 years. It should be noted, that evaluating temporal trends of ΣDDTs in marine mammal populations, and linking them to trends in other biota or environmental media can be confounded by factors that affect tissue concentrations such as age, life span, sex, reproductive activity, and blubber thickness [Arctic Monitoring and Assessment Programme (AMAP) (2004)]. The concentrations in marine mammals also reflect their exposure history over many years, which could result in a significant lag in response to changes in their exposure levels. ΣDDTs concentrations declined significantly in

Table 3.6: Half lives [a], estimated from linear regression in the 1980s.

Location	Organism	$t_{\frac{1}{2}}$ [a] Σ DDTs	
		observed	modelled
Hudson Bay	polar bear (f)	12	9
Arctic Bay, Grise Fjord (Baffin Bay)	ringed seal	14	12
Holman (Beauford Sea)	ringed seal	18	15
Prince Leopold Island	black-legged kittiwake	13	13
	northern fulmar	16	13
	thick-billed murre	17	13

the 1980s in ringed seals from both locations, in the Baffin Bay, and in the Beauford Sea. The respective half lives of 18 and 15 years are close to the ones derived from surface water concentrations, 15 years and 13 years. The contaminant burden in seabird eggs reflects residues assimilated over a long time period by the female, and particularly in migratory species may integrate exposure from a number of different locations. Nonetheless, the half lives derived from egg contamination were compared to half lives derived from seawater concentrations in the ocean adjacent to the location were the eggs were collected. The contamination with ΣDDTs declined significantly in the time period considered for deriving half lives. Half lives of ΣDDTs differ for the three species, with the fasted decline observed in black-legged kittiwake (half life of 13 years) and the slowest in the thick-billed murre (17 years). The surface ocean concentration showed an apparent half live of 13 years. The Arctic Monitoring and Assessment Programme (AMAP) (2004) states that the temporal trends in these seabird egg concentrations reflect changes in contaminant deposition in the marine environments, rather than shifts in trophic level over time, indicated by a stable-nitrogen isotope analyses.

Discrepancies between model prediction and observational data of DDT and DDE The comparison of model data with observations shows significant discrepancies for the spatial distribution in atmosphere and ocean. The observed concentrations show a decrease from low to high latitudes, whereas the model results show an increase of oceanic concentrations. Therefore, modeled concentrations in the Arctic atmosphere strongly exceed observations. In general, possible could be incorrect model results and uncertainties of the observations. Incorrect model results can be caused by uncertainties in the emission data and chemical properties of the compounds, and by incompletely reproduced or lacking relevant environmental processes.

The emission inventory for DDT was built on agricultural usage data reported to the Food and Agriculture Organisation (FAO), and does not include DDT usage for forestry, usage in health programs (e.g. for indoor residual spraying), or storage and disposal as waste [Semeena (2005)]. Only a limited number of countries report to FAO, as reporting to FAO is voluntary [Voldner and Li (1995)]. For these reasons the emission inventory is expected to underestimate the actual sources of DDT. However, no general trend of a underestimation of environmental concentrations in the model is found based on the comparison with observational data. Furthermore, the latitudinal distribution of the emissions is probably less subject to errors than the absolute value. Hence, deficiencies of the emission inventory are not the major cause of deviations between model data and observations.

The range of published water solubility data for DDT spans two orders of magnitude and the true value is unknown due to poor data quality and inadequate documentation of reporting measured and estimated values [Pontollilo and Eganhouse (2001)]. Here a value of $S_w = 0.0034$ mg/L at 296K was chosen. After evaluation of existing data Shen and Wania (2005) suggest to use a value of 0.15 mg/L at 298K. To estimate the impact of differing solubility data, the multimedia mass balance model MPI-MBM was run with two different values, $S_w = 0.0034$ mg/L at 296K and 0.1 mg/L at 291K, which lie within the standard deviation of published data. The model was forced with DDT applications following the same temporal pattern as in the MPI-MCTM model runs. Two different mean surface temperatures were simulated, T_{mean}=287.7 K and T_{mean}=297.7 K, with a constant annual cycle repeated for all model years. The model results show (Figure 3.10) temperature and solubility effects. Larger soil burden is simulated under the lower temperature. This is due to reduced volatilisation from soil, which is temperature dependent. Also volatilisation from ocean is reduced. Differences for the two solubilities propagate most into differences in soil burden, in volatilisation from the ocean, and in oceanic burden. More mass is stored in soil assuming higher water solubility (Figure 3.10). The MPI-MBM model results indicate, that if the solubility of DDT was significantly higher than the chosen one, outgassing of mass from ocean to atmosphere would decrease significantly. Less mass would be recycled from the ocean to the atmosphere in subsequent hops and this way less mass would be transported to the Arctic. At the same time less mass is shifted from soil to ocean, because of reduced volatilisation from soil. Thus, a higher than observed gradient between the tropics and the Arctic could be explained by to low value for water solubility. Because of the superposition of feedbacks affected (transport, cycling in soil and vegetation, deposition and degradation rates) the order of a reduction of the gradient cannot be quantified, even not roughly.

A parameter which determines the atmospheric lifetime of DDT and, hence, long-range transport is the reaction rate coefficient with the OH radicals. In a recent study by Lammel et al (2009) a much higher upper limit for the degradation rate than used in the MPI-MCTM simulation was inferred from DDT observations. Sensitivity towards a higher degradation rate was tested using the MPI-MBM, $k_{OH} = 1.5 \cdot 10^{-12} cm^3 molec^{-1} s^{-1}$ in comparison to $k_{OH} = 1.0 \cdot 10^{-13} cm^3 molec^{-1} s^{-1}$. Again two different mean surface temperatures, $T_{mean}=287.7$ K and $T_{mean}=297.7$ K were simulated. The results expectedly suggest the burden in the atmosphere in is lower for higher degradation rates, and can be expected to be the most pronounced in the northern hemisphere summer. The impact is not linear, because of lifetime in air being still limited by deposition processes rather than degradation. However, deposition to the ocean is reduced and the DDT burden in the ocean is lower under high k_{OH}. As the OH concentration strongly depends on season and latitude and is lower in low latitudes than in high latitudes [Spivakovsky et al (2000)], the atmospheric degradation of DDT in the low latitudes would be enhanced, less mass would be transported via the atmosphere and a reduction of the atmospheric, and, due to reduced deposition, also oceanic concentration in the Arctic could be expected.

The concentration of DDE in air and seawater of the Arctic was shown to be strongly overestimated. As the photolytical degradation rate of $2.0 \cdot 10^{-11} cm^3 molec^{-1} s^{-1}$ used in the model simulation is much higher than the upper limit suggested by Lammel et al (2009) of $0.75 \cdot 10^{-12} cm^3 molec^{-1} s^{-1}$, an underestimation of the degradation of DDE in the gas phase can not be the reason for these discrepancies. Generally an insufficient representation of the metabolisation of DDT into DDE, a missing sink process, or an erroneous volatilisation from soil should be considered as reasons for the discrepancies. The parameterisation of volatilisation from soil in the model is indeed insufficiently validated for DDT/DDE. An overestimation of the volatilisation could contribute the high levels of DDT and DDE in the Arctic air and ocean. On the other hand the satisfying agreement of DDE concentrations in seawater of the Bay of Bengal and the South China Sea, despite erroneous air concentrations, indicates that an insufficiently represented volatilisation from soil alone can not explain the discrepancies between model results and observations.

Processes not resolved in MPI-MCTM that also might serve as explanations for discrepancies between model results and observations are degradation of DDT in the ocean and degradation of DDT in the particle-bound state in air. Furthermore, the neglection of sea-ice had been shown to increase the northern hemisphere meridional DDT gradient [Guglielmo (2008)].

Particle phase reactions of pesticides in the atmosphere is are an area of great uncertainty [Atkinson et al (1999)], and no direct conclusions about possible impacts can be drawn from just the fact that they are not resolved in the model. High particle bound mass fractions are predicted in high latitudes (>80 %) in winter. Thus, degradation in air, as it is assumed to be limited to the gaseous phase, is reduced. An additional degradation process in the particle phase is assumed to reduce concentrations in the Arctic, consequently. On the other hand lifetimes of particle-bound DDT is limited by deposition, much more than in the gas-phase.

The distribution of DDT and related compounds in fish suggests that they can convert DDT to DDD and DDE, which has been confirmed by studies in which fish of various species have been fed with DDT or have accumulated it from water [Ernst and Goerke (1974), Addison and Willis (1978),Leah et al (1997)]. This sink process would reduce the level of DDT in sea water and would imply an additional source of DDE, not considered in the model simulation. Despite degradation in fish also degradation by marine microorganisms in the water column [Patil et al (1972)] is not represented in the model. Its significance, however, should be limited to the productive sea regions, i.e. shelf seas and upwelling regions.

The comparison of the reduction rate of ΣDDT concentration in surface sea water in the 1980s with the one in several Arctic organisms (Figure 3.6) however suggests, that the oceanic reduction rate is well represented in the Arctic ocean, as the half lives derived from model predictions do not differ significantly from the ones found in Arctic biota. Although a comparison of trends in Artic biota with trends in seawater is not straightforeward, as it is impacted by factors such as migration behaviour of the organism, it can be used as an indicator for the reduction rate in the ocean. If the observed decline would have been much faster than the one predicted by the model, this fact could have been interpreted as an indicator for a lacking sink process in seawater. But, as observed and predicted half lives are in good agreement, a lacking oceanic sink is not believed to be a decisive reason for the discrepencies between modeled and observed concentrations of DDT and DDE in the Arctic ocean.

3.1.4 Summary

Global usage of DDT and its breakdown to DDE in were simulated for 1950-1990. The model results suggest that the global environmental contamination peaked in all compartments with a wide range of delay, 1-17 years compared to the emissions. For the first time a geographically resolved distribution of historic years of peaking DDT contamination was addressed. Total environmental burden in individual regions is shown to reach maximal contamination levels generally 5-10 years after the application maximum. Although also the ocean has been shown to loose mass since 1977 some oceanic regions, like the Arctic and Tropical Pacific did not peak up to 1990, but are continuously accumulating mass until the end of the simulation. Also the deep ocean accumulated mass until the end of the simulation. Residence times of DDT in the ocean is largest in the Arctic ocean, where volatilisation is reduced due to low surface temperatures. The total environmental residence time of DDT was estimated to 16 a.

The comparison of model results with observations reveals significant discrepancies, especially in the Arctic, where concentrations are overestimated by the model in ocean and atmosphere. The model does not reproduce the observed latitudinal gradient of DDT in the ocean, i.e. the modelled and observed gradients have a different sign. Observations show, that the contamination of the ocean by DDT is decreasing towards the Arctic, whereas in the model simulation contaminations are increasing.

Due to the complexity of the model the prospect to identify a single processes responsible for discrepancies between model results and observations is low. And a comprehensive sensitivity analysis is not feasible due to computing costs. The possible sensitivity to selected input parameters was addressed by MBM simulations. It could be shown that some substance properties, i.e. the photochemical degradation rate and solubility of DDT, have the potential of reducing the erroneous latitudinal gradients in ocean and atmosphere. Also the implementation of processes not yet captured, like degradation of DDT in the particle phase in air or degradation of DDT in the ocean could contribute to better model results. The ongoing increase of DDE in soils and ocean stands in contrast with observations and indicates the existence of other efficient environmental sinks of DDE, other than reaction with OH radicals in the atmosphere and dechlorination to DDMU. The photolytical degradation rate used in the model simulation is much higher than the upper limit derived by Lammel et al (2009). Air-soil exchange is one process not sufficiently validated for semivolatile organic compounds, that was assumed to be a reason for erroneous model results.

3.2 Environmental fate and distribution of perfluoroctanoic acid

3.2.1 Introduction

Since the 1950s perfluoroctanoic acid (PFOA) and its ammonium salt (APFO) have been used as processing aids in the manufacture of fluoropolymers like polytetrafluoroethylene (PTFE, commonly known as Teflon and Gore-Tex). Because of its water and grease repellent properties, it is used in a wide range of consumer products such as cookware, paper, fabrics and carpets. PFOA is highly persistent in the environment and biota with reported half-lives of several years in humans [Olsen et al (2007)]. It has been observed in low levels in the serum and liver of unoccupationally exposed individuals and various wildlife species at a number of geographic locations [a.o. Olsen et al (2003), Senthilkumar et al (2007), Falandysz et al (2007)]. Levels in urban environments are likely to result from point sources, whereas levels found in remote areas, due to the compounds low atmospheric mobility, are likely to result from degradation of mobile precursors like fluorotelomer alcohols (FTOHs) [Ellis et al (2001), Ellis et al (2004)], and oceanic transport.

Toxicity and exposure studies indicate PFOA is immunosuppressive and can cause developmental problems and other adverse effects in laboratory animals, such as rodents [Lau et al (2004), Lau et al (2006)]. In 2005 the US Environmental Protection Agency (EPA) released a draft risk assessment of its potential human health effects [U.S. EPA (2005)]. A subsequent review by the EPA science advisory board concluded that there is sufficient evidence to classify PFOA as likely human carcinogenic.

For its high persistence and widespread detection in various environmental matrices, PFOA arouse large scientific interest in recent years. A number of box model studies were conducted to estimate its transport pathways in remote regions via oceanic currents and degradation of precursor substances [Prevedouros et al (2006), Armitage et al (2006), Wallington et al (2006), Schenker et al (2008b)]. The current study is the first attempt to identify oceanic transport pathways using a comprehensive model with high spatial resolution. Its results are compared to observed PFOA concentrations.

PFOA physical and chemical properties PFOA is a completely fluorinated organic acid. Due to the strength of the high-energy carbon-fluorine bond and effective shielding of carbon by fluorine atoms, PFOA is exceptionally stable against metabolic or environmental degradation (hydrolysis, photolysis, microbial degradation) [Kissa (2001), Giesy and Kannan (2002)]. The only potential natural degradation process is gaseous PFOA reacting with OH radicals in the atmosphere. Hurley et al (2004) analysed the reaction of several gas-phase perfluorinated carboxylic acids (PFCAs) with OH radicals and their resulting atmospheric lifetimes. Their findings imply longer atmospheric lifetime for longerchainPFCAs. Therefore, PFOA is expected to be very persistent in the environment.

In addition to degradation, partitioning to organic matter and intermedia mass exchange define the fate of a substance in the environment. PFOA has a dissociation constant pK_a of 2-3 [Brace (1962)], hence it will be present in natural waters predominantly in its dissociated form, that is, as perfluooctanoate (PFO). The anionic PFO is not expected to partition into the gas phase due to its low vapour pressure, and consequently remains in water. The fluorinated hydrophobic tail, which is lipophilic, and hydrophilic carboxylate head group of PFO determine its behaviour at interfaces. Due to their amphiphilic structure PFO molecules associate to form micelles in aqueous solutions. In a micelle the hydrophilic head group is exposed to the surrounding water and the hydrophobic tail makes up the interior of the micelle. Because of its surfactant properties, PFO will preferentially adsorb to the water surface. Since PFOA tends to form multiple layers, K_{ow} (octanol-water partition coefficient) cannot be measured for PFOA [U.S. EPA (2005)]. Partitioning to organic matter can also be expressed using the partition coefficient to organic carbon. Arp et al (2006) found that common methods like the Junge and Finizio model to predict the gas-particle partictioning underestimate the amound of PFCa bound to organic matter. The actual, efficient sorption to an organic phase dispersed in water will exceed the sorption to the same phase not dispersed in water, because of the amphiphilic nature [Boethling and Mackay (2000)]. The exceedence will be a function of the surface-to-volume ratio of the interface. To estimate the impact of partitioning to organic matter therefore two different sorption coefficients, deviating by a factor of 100 were tested. A minimum K_{oc} of 115 mg/L measured by Higgins and Luthy (2006) and a maximum of 11500 mg/L were chosen.

A list of properties relevant for multimedia fate modelling is given in Table 3.7.

PFOA emissions PFOA emissions arise from direct and indirect sources. The term ' direct sources' refers to PFOA released to the environment during the manufacturing and use of PFCAs. Degradation from precursor substances, like fluorotelomer-

Table 3.7: Physico-chemical properties of PFOA.

Physico-chemical property	Value
Saturation vapour pressure [Pa]	4.2 (298 K)
Enthalpy of vaporisation [kJ/mol]	95
Water solubility [mg/L]	4100 (295 K)[a]
Enthalpy of solution [kJ/mol]	2.5[b]
Henry coefficient [Pa m^3/mol]	0.1831 (298K)
Octanol-air partitioning coefficient (logK_{oa})	6.8 [c]
Soil organic carbon-water partitioning coefficient (K_{oc})	115-11500
OH gas -phase rate constants[cm^3/moles/s]	$1.69 \cdot 10^{-13}$ [d]
Acid-ionisation constant (dissociation constant) pK_a	2.8[e]
Degradation rate in ocean [1/s]	0.00

[a] Prevedouros et al (2006)
[b] Tomasic et al (1995)
[c] Arp et al (2006)
[d] Hurley et al (2004)
[e] Brace (1962)

alcohols (FTOHs), is an example of an indirect source.

PFOA is released in the manufacture of its ammonium salt (APFO). It was produced from 1947-2002 mainly by electrochemical fluorination [Prevedouros et al (2006)]. The largest production sites were situated in the United States, Belgium, Italy and Japan. The estimated global emissions from the manufacture of APFO from 1951-2004 are approximately 400-700 t. APFO is a processing aid in manufacture and processing of fluoropolymers, facilitating the aqueous polymerisation of fluoromonomers. The historical global emissions from fluoropolymer manufacture for the years 1951-2003 range from 2400 to 5400 t. A recent trend shows a decrease in APFO emissions, because of the substitution of APFO by other processing aids. Other direct sources of PFOA are fluoropolymer dispersions, aqueous fire-fighting foams and consumer and industrial products.

Prevedouros et al (2006) estimate PFCA point source emissions from fluoropolymer manufacturing to contribute with approximately 60 % to the total PFCAs used. The substance distribution upon entry is 23 % to air, 65% to water and 12 % to soils [Prevedouros et al (2006)].

For the emission inventory used in the model experiments input data from Armitage et al (2006) for the years 1950-2005 (Figure 3.12b) were compiled into point source emissions on the model grid. Armitage et al (2006) built their emissions upon the aforementioned review of Prevedouros et al (2006) including only direct sources in their emissions. In the model the amount emitted directly to the atmosphere and land compartments is distributed equally to four gridcells located in the countries of the largest production sites (United States, Belgium, Italy and Japan) (Figure 3.12a). All point sources follow the same temporal emission evolution, starting in 1950. For Italy, Belgium and Japan the emissions to the ocean compartment are introduced into the adjacent seas, which are the Adriatic Sea, the North Sea and the Pacific

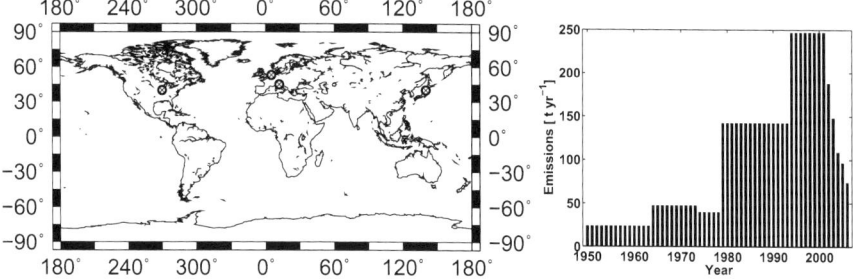

Fig. 3.12: Locations of the point source emissions in the model (left). Annual PFOA emissions [t] adopted from Armitage et al (2006)(right).

Ocean (Tokyo Bay), respectively. The emission source for the U.S. is located in the Midwest, hence the closest water body is the Great Lakes. As the Great Lakes are not represented in the model in terms of a water compartment for chemicals cycling, the emitted mass is assumed to pass trough the Great Lakes and enter the Atlantic Ocean via the St. Lawrence River. Boulanger and Peck (2005) show volatilisation of PFOA in Lake Ontario is less than 1% of the inflow. Therefore we adopt a 1 % loss rate, neglecting any other loss than volatilisation. Assuming the same loss rate for Lake Erie, Lake Michigan, Lake Huron and the St.Lawrence River the emissions in the USA to the North Atlantic are weighted by a factor of 0.99^5 and placed into a surface gridcell of the ocean model at the mouth the of St. Lawrence River, Canada. The volatilisation losses are added to the direct emissions into air at the Midwest USA source point.

PFOA observations To evaluate MPI-MCTM model results observational data of PFOA from ship cruises in the Atlantic, Indian and Pacific Oceans were taken from literature (summarised in Yamashita et al (2008)). The data was collected between 2002 and 2006 in a global ocean monitoring initiative. Samples were taken from ocean surface water. Vertical profiles were sampled in the Labrador sea, the Mid Atlantic ocean, the South Pacific ocean and the Japanese sea, where water probes were done at several depths down to 5500 m. The limit of quantification for PFOA was determined as 6 pg/L.

3.2.2 Experimental setup

Due to the fact that PFOA is chemically stable and highly persistitent in the environment it is necessary to get environmental background contamination for experiments starting later than 1950. To reduce computational costs MPI-MCTM was run for forty years in a coarse resolution of T21GR30 to create realistic background values for a model run starting 1991. Two PFOA tracers with different sorption coefficients to organic carbon, KOC=115 mL/g (KOC115) and KOC=11500 mL/g (KOC115) were simulated. In the case of KOC115 248.784 t were degraded within these 40 years and in the case KOC11500 182.54 t. The degraded mass was subtracted from the total mass applied leading to an environmental burden of 2271.216 t (KOC115) and 2337.46 t (KOC11500). At the end of the simulation over 90% of the total mass ended up in the ocean, followed by 1% (KOC115) and 5% (KOC11500) in soil. Consequently, pollutant masses were distributed among the compartments following the mass distribution at the end of the coarse resolution run. MPI-MCTM was run from 1991-2004 in T63GR15 resolution using the aggregation module for particle sinking in the ocean.

Dissociation of the neutral acid in water necessitates modifications for air-sea exchange in the model, which is based on Henry's law. Other possible pathways, e.g. sea spray, are neglected. Henry's law is restricted to concentrations of physically solved, non dissociated substances. Since only the non-dissociated acid is volatile, it is important to correct the air-water partition coefficient as to reflect the relative proportions of volatile and non-volatile components. The corrected parameter is the effective Henry's law coefficient, which is related to the Henry's law coefficient as a function of pH (modified Henderson-Hasselbalch equation):

$$H_{\text{eff}} = H/(1 + \frac{K_a}{[H^+]}) \qquad (3.1)$$

$[H^+] = 10^{-8.1}$ at pH=8.1, K_a is the dissociation constant, H is the pure component air-water partition coefficient.

3.2.3 Results and discussion

Environmental distribution The temporal evolution of global PFOA burdens in different environmental compartments (atmosphere, ocean, vegetation and soil) is shown in Figure 3.13. For both tracers, KOC115 and KOC11500, the majority of mass is found in the ocean. At the end of the simulation, the ocean holds over 97 % of the environmental burden (Table 3.8). The oceanic burden does not show any pronounced annual cycle, because its primary source, direct emission, is constant over the year and two orders of magnitude larger than deposition from the atmosphere. 12% of the mass were emitted to soil and 23 % to atmosphere, whereas only 0.2-0.3 % end up stored in atmosphere and 0.6-2.7 % in soil. Differences between the KOC115 and KOC11500 tracers are caused by its different lipophilicity. A higher partitioning coefficient to organic carbon increases the fraction irreversibly bound to organic matter in soil for equal environmental conditions. Higher mass levels in soil for the more lipophilic tracer KOC11500 are balanced by decreased burdens in the ocean. Vegetation encounters the lowest PFOA burden as no mass is directly emitted to vegetation, the lipophilic character of this compartment is neglected in the parameterisation, and the only source for this compartment is deposition from atmosphere. In addition, PFOA does not undergo any degradation in soil and vegetation, hence volatilisation is the only sink for these compartments. The annual cycle in both compartments originates from seasonal variations of deposition and volatilisation. Volatilisation is temperature dependent and high in warm seasons. Deposition processes, both dry and wet, are also influenced by the occurrence of precipitation events and seasonal variations of partitioning between gas and particulate phase. The geographical distribution of the sources, located exclusively in the northern hemisphere, suggests that both processes attain its maxima in northern summer. The model results show volatilisation maxima from soil and vegetation in August. Dry deposition also peaks in August, but wet deposition rates are largest in October. Due to its molecular structure PFOA is very persistent in the environment.

Table 3.8: Compartmental distribution at the end of the simulation [%] and residence times [a].

Compartment	Mass fraction [%]		τ [d]	
	KOC115	KOC11500	KOC115	KOC11500
Atmosphere	0.20	0.26	57.75	58.40
Soil	0.63	2.71	153.30	584.00
Vegetation	<0.001	0.001	2.56	2.56
Ocean	99.16	97.03	147 715 50	147 861 5
Total	100.00	100.00	401.30 a	393.11 a

The only sink process simulated by MPI-MTCM is degradation of gaseous PFOA in the atmosphere. The overall residence time estimated from model results is 401.30 a for KOC115 and 393.11 a for KOC11500 (Table 3.8), differing for the two tracers because of less mass available for degradation in case of KOC11500 (evident in Figure 3.13). Residence times vary for different compartments and are highest in ocean, caused by a high environmental burden and a low volatilisation tendency of PFO. The residence time in the atmosphere is approximately 58 days (Figure 3.8). Atmospheric lifetimes of gaseous PFCAs have been previously examined [Hurley et al (2004)], and estimated to a few days to several weeks. This study further indicates that the atmospheric residence time of PFOA is limited by dry and wet deposition rather than by reaction with OH radicals. In the model simulation 30 t are removed per year by reaction of gaseous PFOA with OH radicals, while deposition removes more, approximately 80 t per year.

The oceanic burden in December 2004 shows the contamination of the ocean after 50 years of PFOA emissions (Figure 3.14). Highest PFOA burden is located in the northern Atlantic, Mediterranean, and the Arctic ocean. Contaminations of the Atlantic, Mediterranean and Pacific can be related to the vicinity to the oceanic source. PFOA in remote regions, however, such as in the Arctic must have been transported via atmosphere or ocean. MPI-MTCM does not simulate degradation of PFOA from volatile, highly mobile precursor substances, that contribute to the ocean burden in the Arctic by deposition. Then annual dry and wet deposition rates of PFOA in the model are small compared to the mass emitted directly to the ocean. This implies that the burden in the Arctic is results mainly from oceanic long-range transport.

To asses the annual influx of PFOA to the Arctic ocean, depth integrated mass fluxes over virtual planes crossing the Norwegian Sea, Denmark Strait, Baffin Bay (Davis Strait) and the Bering Strait were calculated. Inflow is dominated by transport with the North Atlantic current via the Norwegian Sea (Figure 3.15), contributing up to 80% to the total transport, which is approximately 12-30 tonnes per year. The second largest inflow comes through the Bering strait reaching 0.7-1.5 t/a. Outflow out of the Arctic ocean takes place in the Denmark strait (4-10 t/a) and Davis Strait (0.1-1.7 t/a). The resulting netflux is positive (import into the Arctic ocean), and varies

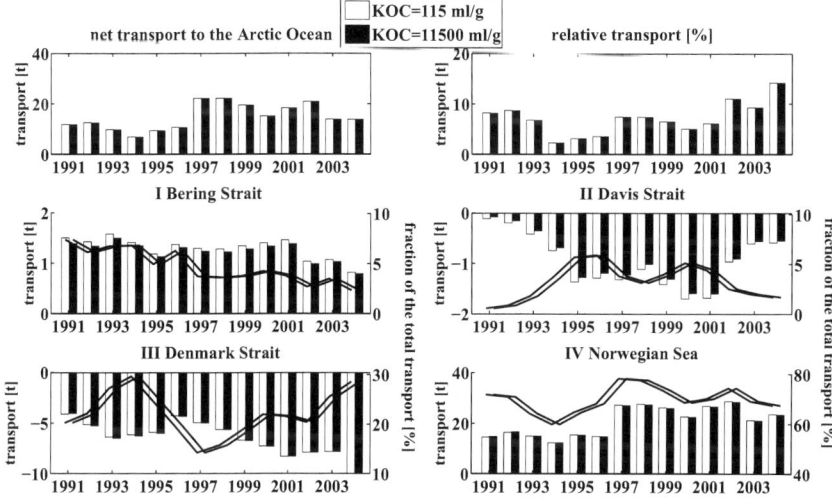

Fig. 3.15: Transport of PFOA to the Arctic ocean, bars: transported mass per year [t], import defined positive, lines: fractions of total transport [%]. The relative transport is derived from dividing the net transport by the emitted mass of the respective year.

between 8 and 23 t/a. The interannual changes do not covary with the temporal pattern of the emissions, but are defined by the variations of the individual currents. This is also evident from by the relative transport per year, which is not decreasing at the end of the simulations in contrast to the emissions (compare Figure 3.15 and Figure 3.12). The low influx 1994, for example, is caused by low inflow to the Norwegian Sea, together with a rather high outflow via the Denmark and Davis Straits. The North Atlantic current system, as part of the thermohaline circulation, is largely driven by temperature and salinity changes, in addition to wind. Therefore, variations in PFOA inflow are likely to be controlled by large scale atmospheric variabilities, such as the North Atlantic Oscillation [Mork and Blindheim (2000)] and changes in freshwater influx (e.g. due to ice dynamics).

The vertical distribution of pollutants originates from water movements and sinking of organic particulate matter operating as carriers for lipophilic substances. PFOA enters ocean water exclusively in the surface level and instantaneously establishes equilibrium with abundant organic phases. The fraction bound to organic matter varies spatially and temporally, depending on organic matter abundance, which in turn is defined by parameters such as nutrient availability and insolation. In June 2004 largest fractions correspond with regions of high primary production due to upwelling of cold, nutrient rich water. These regions are in the equatorial Pacific, and in areas of coastal upwelling (Figure 3.16). Regions of coastal upwelling are the Canary Current (Atlantic, 10-40 ° N), the California Current (Eastern Pacific, 25-40 ° N) and the Benguela Current (Atlantic, 10-30 ° S). These maxima of mass bound to organic matter are spread in the uppermost 20 m, decreasing rapidly until 80 m depth (Figure 3.17). This can be seen for both tracers, however, for KOC115 values are 100 times lower than for KOC11500. In general the highest abundances of PFOA in organic phase are found in areas with low absolute PFOA burden (evident from a comparison of Figure 3.16 and Figure 3.14). From a vertical and latitudinal distribution it is evident that most PFOA mass is located in depths of 300 to 500 m, with a prominent wedge of high mass penetrating downward from the equator near the surface to about 40 ° at 2500 m depth. Although concentrations are at a much smaller scale, a significant amount of mass is found in the southern hemisphere. At about 60 ° S traces of PFOA can be detected down to 5000 m. The amount of PFOA stored in 60-70 °N is of the same order of magnitude as in the southern ocean, but concentrations are much higher in the Northern hemisphere (Figure 3.14). The vertical stratification of PFOA can be related to corresponding sinking of dense water. Most of the sinking water is formed by cooling in mid to high latitudes, although some is form by evaporation, as for example in the Mediterranean Sea [Peixoto and Oort (1992)]. Deep convection in the European Arctic produces water masses descending to the bottom of the North Atlantic after flowing over sills between Greenland and Scotland. Water masses formed by convection in the Labrador Sea do not penetrate down to the bottom, but reach only intermediate depths [Mann and Lazier (2006)]. Deep water formation by convection also occurs over shelfs around Antarctica, as in the Weddel Sea. This probably explains PFOA descending to depths of 5000 m around 60 ° S. From the difference between the simulated December of 1991 and 2004 an annual flux of PFOA below 100 m depths of 145 t/a can be estimated.

Comparison with observations   To evaluate model results a comparison to observational data from several ship cruises was conducted. Merged data from 2002-2006 are compared to annual mean concentrations in 2004. This is feasible, because the spatial pattern of the model results does not change significantly in the last 4 years of the simulation, due to invariant emissions. A large difference between the northern and southern hemisphere's contamination can be seen in both, observations and model results. Surface waters in the northern and mid Atlantic Ocean and northern Pacific are higher contaminated with PFOA than waters in the southern hemisphere. Observations show largest surface water concentration in the mid Atlantic Ocean and in locations east of Japan. In the model the distribution of PFOA in surface waters is determined by the emission scenario. Contaminations are highest in areas close to the point source . Large concentrations in the South China sea, south of Japan and in the mid Atlantic, south of the point sources are not captured. The

American point source of PFOA in the model is situated in the Midwest, and discharge to the ocean is assumed only via the St. Lawrence river. A dominant global producer of PFOA is the Minnesota Mining and Manufacturing Company (3M). It has two plants in the U.S, in Decatur, Alabama and Cottage Grove, Minnesota, were fluorochemicals are produced. The Minnesota Pollution Control Agency estimated an annual discharge of 23 000 kg of PFCs from the 3M wastewater treatment plant into the Missippi river [Oliaei et al (2006)]. Their measurements downstream of the river revealed PFOA concentrations of 35 ppt. As PFOA is very persistent and not volatile, discharge from Mississippi is assumed to be one source of PFOA to the middle Atlantic Ocean via the Gulf Mexico, which is not captured by the model.

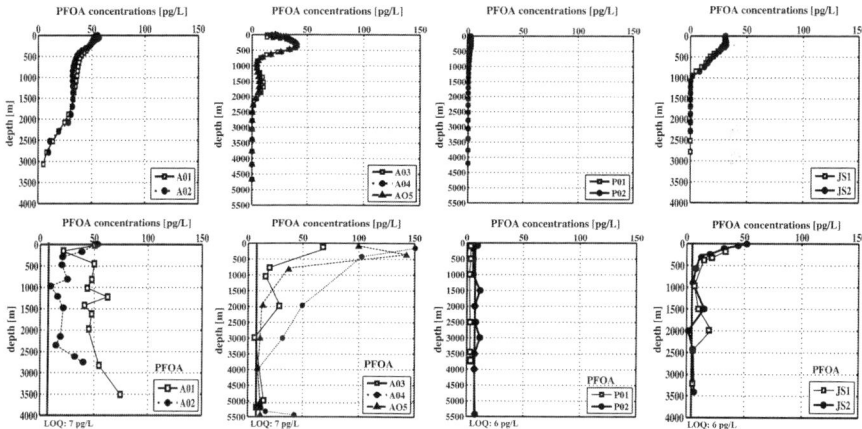

Fig. 3.16: Vertical profiles PFOA concentration [pg/L] in the Labrador Sea (A01,A02), Mid Atlantic Ocean (A03, A04, A05), Japanese Sea (JS1, JS2) and South Pacific Ocean (P01, P02); upper panels model results, lower panels Yamashita et al (2008) a) September 2004, AO1 56° 34' N 52 ° 48' W, AO2 56 ° 41' N 39 ° 40' W b) March 2004, AO3 23 ° 17' N 64 ° 19' W, AO4 25 ° 47' N 64 ° 59' W, AO5 27 ° 03' N 64 ° 35' W c) December 2004, PO1 67 ° 12' S 169 ° 57' W PO2 39 ° 59' S 169 ° 59' W d) May 2005, JS1 40 ° 43' N 136 ° 24' E, JS2 44 ° 12' N 138 ° 54' E .

Vertical profiles from different ocean regions differ significantly from each other. In the Labrador Sea (Figure 3.16a) PFOA concentrations are 50 pg/L at the surface for both model results and observations. For AO1 and AO2 modelled profiles are almost identical, while observed profiles behave differently. Concentrations in water sample at AO1 are relatively constant throughout depth, except for subsurface water, where PFOA concentration decreases, and water below 2000 m in which concentrations increase. Modelled concentrations, as well as observed ones at AO2, decrease until 500 m, and remain constant down to 2000 m. In waters below 2000 m PFOA concentration increases for observations, but decreases in the model results. Yamashita et al (2008) suggest that water masses from the surface down to 2000 m were well mixed due their convective formation. The subsurface is explained by a decrease by an influx of the Labrador Current containing melt-water, and the increase in larger depths by the presence of an independent deep water current carrying higher amounts of PFOA. A similar stratification is observed in the model results, except for the deep water current, that decreases concentrations in waters below 2000 m.

In the mid Atlantic ocean observed concentrations at AO3 and AO4 decrease gradually with depth, whereas AO5 concentrations show an increase until 500 m and a decrease below. Modeled profiles show a similar pattern. Surface concentrations of model results are much lower than the observed ones. In addition to this the fact that profiles of all mid Atlantic sampling location are identical can be explained by missing discharge into the mid Atlantic ocean in the emissions scenario. Emissions from American fluoropolymer productions sites are released into the Atlantic Ocean solely at the mouth of St. Lawrence River. Discharge of PFOA into for example the Gulf of Mexico is not considered.

Profiles in the Japanese Sea are similar for model and observational data. Concentration decreases down to 1000 m and remains constant below. Surface concentrations are lower for modeled profiles, most likely due to the emission scenario, that assumes identical temporal behaviour for all source points and does not capture all emitted mass. Due to the limited horizontal resolution of models, the topography of the ocean differs from the real one. In the Southern ocean concentrations were low throughout all depths, and for the measurements often below the detection limit of 6 pg/L.

3.2.4 Summary

Over 97 % of the emitted mass of PFOA ends up in the oceanic compartment. Armitage et al (2006) predicted over 80 % of PFOA in the ocean, regardless of the model scenario. In the ocean PFOA gets transported with oceanic currents and reaches remote areas as Arctic and Antarctic Oceans. Although almost not bioaccumulative [Conder et al (2008)] and only in rather small amounts bioavailable via trophic transfer its abundance in sea water makes it available for aquatic organisms. Simulated fluxes to the Arctic ocean are of the same order of magnitude as estimated in other studies [Prevedouros et al (2006), Armitage et al (2006)].A longitudinal distribution of transport to the Arctic is addresses for the first time. Variability of the import of PFOA to the Arctic was found high, and mainly limited by the variability of oceanic currents, rather than the temporal emission pattern. Two PFOA tracers of different lipophilicity were modelled. The differing sorption to organic carbon results in a diverging compartmental distribution. The more lipophilic KOC11500 is found in higher fractions in soil, which increases its persistence. Oceanic vertical and horizontal distribution is defined by currents and due to only low fraction bound to organic matter differences between KOC115 and KOC11500 are very small.

Although small in concentration units a significant amount of mass was found in southern hemisphere oceans.

Model results in seawater are in good agreement with observational data of PFOA. Most differences can be attpageributed to deficiencies of the emission scenario. Despite this fact, the difference between model results and observational data are due to the limited horizontal and process resolution and the fact that the physical parameters of the model (temperature, surface pressure, vorticity or divergence of the wind velocity field) were not relaxed to observational data. Regarding these limitations, in particular individual vertical profiles compare quite well with observations. This study underlines the importance of the ocean as a transport medium of PFOA. The contribution of volatile precursor substances to long-range transport needs to be assessed.

Chapter 4
Conclusions and outlook

In this study the role of the ocean in global cycling of persistent organic contaminant was examined. The long-term simulation of DDT showed that the ocean only partly acts as a final sink of semivolatile organic pollutants (SOCs), which are highly persistent in the ocean. Its capacity to permanently store these pollutants is highly spatially variable. In the warm tropical oceans volatilisation forms a major part of the cycling. Residence times of persistent SOCs are low in the tropics, because most of the pollutant mass is found in the surface layers. The Arctic Ocean, on the other hand, represents the region where persistent SOCs are stored longest, mainly due to reduced volatilisation in combination with high deposition. Dynamical and particle bound transfer of semivolatile organic pollutants is of high importance for the fate of pollutants, because, as shown in Chapter 3.1 for DDT, only the deep ocean accumulates mass over a long period of time, whereas the surface ocean has only a limited storage capacity. The transfer of lipophilic organic pollutants, such as DDT into the deep ocean by gravitational settling of particulate organic matter is sensitive to size and sinking speed of the particles. Large particles with high sinking speeds are especially important for the transfer of pollutants from sub-surface waters to the ocean bottom. Aggregation of large marine snow particles is hence an important process for cycling of lipophilic organic contaminants. Although the sensitivity of LRT of DDT to the different representations of suspended organic matter in the ocean was found to be low, the long-term simulation of DDT shows that on longer time scales the amount of mass transfered into the deep ocean is decisive for the compartmental distribution of the pollutant. Hence, for the long-term fate of persistent lipophilic SOCs size and sinking speed of marine organic particles is potentially a controlling parameter for long-range transport.

In particular with regard to the role of marine organic matter in global cycling of persistent organic contaminants, immediate questions follow this study. The individual impacts of particle size dependent sinking versus a high gradient of organic matter concentrations on the ocean margins on pollutant cycling were not isolated. A 'control' experiment, using only the higher model resolution and assuming a 5 m/d constant sinking velocity of POC, could be used to assess the contributions separately. This experiment might also include an additional tracer, which is not partitioning to organic matter, but is fully dissolved in sea water. This tracer could then be used to quantify the impact of organic matter as opposed to dynamic sinking, i.e. downwelling or deep-water formation, without diagnosing it from an indirect measure , i.e. export production.

For global modelling of lipophilic organic compounds the implementation of aggregation of marine snow is important to realistically simulate biogenic tracer distributions. The model, however, only partly captures implications of the aggregation of marine snow on pollutant fate. Partitioning to organic matter in the model is parameterised as a function of organic matter concentration, and is derived from the organic carbon partitioning coefficient [Schwarzenbach et al (2003), Guglielmo (2008)]. Studies in marine PCB cycling [Cziudaj (2005)], however, indicate, that the pollutant concentration of the sinking organic matter decreases with increasing depths. Cziudaj (2005) assumes that this caused by high sinking velocities of particles and their transformation processes during the sedimentation in the deep ocean. As the biogeochemistry model HAMOCC5 resolves a size distribution with varying sinking speeds of particulate organic matter, further studies on the impact of sinking speed and particle size on partitioning of pollutants could help to improve the representation of pollutant transfer by gravitational sinking in the model, and, hence, understand the role of gravitational settling on a global scale.

Due to the high biological activity in continental shelves, contamination of shelf regions by bioaccumulating pollutants poses a main hazard to the marine environment. Due to their dynamic characteristics, continental shelves play a special role in the cycling of pollutants in the ocean. Tidal movements and storms can lead to a release of contaminants stored in the sediments [Eggleton and Thomas (2004)] remobilising the contaminants.Wiberg and Harris (2002) showed, that 25-50 % of the pollutant mass (p,p'-DDE) in the surface active layer of the bed is lost during a resuspension event. Also anthropogenic activities, including maintenance and capital dredging, and the disposal of historically contaminated estuarine sediments, result in major sediment disturbances. Transport of pollutants into the shelf sediment is sensitive to the concentration of suspended matter in the shelf regions. Hence a good representation of biogenic tracers in continental shelf zones is important.

The assimilation of satellite data presented in the this study only partly improved the the representation of continental shelves in the global model. To achieve a better representation, in future simulations either the model simulation should last for a longer time, the assimilation should be stronger or additional processes which may lead to an increased organic matter concentration in the shelf zones, such as a nutrient entry from rivers, need to be included. In order to study the resuspension of pollutants from the sediment, sediment would need to be included as a compartment for chemical cycling into the model, e.g. as a two phase system (including a particulate organic phase and pore water). The sediment compartment could make use of the sediment description of HAMOCC5 [Maier-Reimer et al (2005)], which includes 12 layers with decreasing porosity.

In Chapter 2.5 volatilisation of DDT from the ocean surface was shown to be disaggregated in regimes of wind speed and sea surface temperature control. This novel insight into the cycling of SOCs discourages the use of zonally averaging models to study e.g. meridional transport of organics: volatilisation from the ocean surface cannot be parameterised consistently. The underestimation of volatilisation rate and hence long-range transport of SOCs by zonally averaging models implies, that the amount of mass transported to remote regions predicted by these models is underestimated. Hence, the use of these model to examine the potential of SOCs produced in middle or low latitudes to contaminate the Arctic is highly not recommendable. The high zonal variation of volatilisation rate furthermore suggests, that also intercomparison of transport behaviour of substances with different physico-chemical properties and spatial emission patterns based on model results of MBMs resolving only zonal bands in order to asses their long-range transport potential is not feasible. More general, approaches in environmental risk analyses to base LRTP and overall persistence on substance properties only, largely or totally neglecting the variabilities of environmental conditions in time and space [e.g. Scheringer (1996), Gramatica and Papa (2007)] are challenged by these results. LRTP and overall persistence principally result from a combination of substance properties and environmental conditions specified in time and space

For PFOA the ocean represents the main repository and an important transport medium. Transport of PFOA into the Arctic is mainly controlled by the variability of oceanic currents, and not directly coupled to the emission pattern. To realistically simulate transport of PFOA, and other amphiphilic substances, more knowledge about their partition behaviour is needed, in particular their surface active behaviour in the aqueous system. The approach of bridging this gap for model simulations by simulating two tracers with differing partition coefficients to organic carbon leads to differing results in long-range transport and compartmental distribution, and hence should be preserved in further studies, until the surface active behaviour can be parameterised differently. The chosen factor of 100 between the partition coefficients was not validated and needs further examination. Direct emission from a few point sources and oceanic transport explain main features of the global distribution of PFOA. Yet, not all features are captured by the model study, and for further investigation of PFOA the emission inventory needs to be improved. Additionally, indirect sources, neglected in this study, need to be considered, in particular formation of PFOA in the atmosphere by oxidation of precursor substances such as fluorotelomer alcohols (FTOHs) [Ellis et al (2001), Ellis et al (2004)]. Due to their volatility these substances may contribute significantly to transport of PFOA into remote regions, as indicated by model studies [Schenker et al (2008b)]. An implementation of a simplified oxidation mechanism, as presented in Wallington et al (2006) or Schenker et al (2008b) into ECHAM-HAM, is needed to address the individual impacts of transport by oceanic currents and atmospheric transport of precursors with subsequent deposition of PFOA, and, hence comprehensively study PFOA cycling in the total environment on the global scale.

For DDT and DDE the ocean plays different roles in global fate of the substances. For DDT the ocean acts both a sink in deep oceanic layers, and as a source to the atmosphere from the surface layers. In case of DDE on the otherhand, the cropland pesticide application memory leads to a strong delay between DDT application and release of DDE, which in turn leads to a delayed contamination of the ocean. Thus, the ocean continues to act as only a sink of DDE, when it already started to act as a secondary source of DDT. The application memory furthermore implies that especially northern regions of the American and Eurasian continent and the Arctic ocean will continue to be contaminated with DDE, when DDT contamination significantly dropped. Neighboring regions can thereby experince a totally different evolution of the contamination. From the comparison of simulated DDT and DDE concentrations with observation it became obvious that MPI-MCTM is not able to capture all features of the global distribution of these two substances. Particular strong deviations were found in the ocean, where the modelled and observed gradient between high and low latitudes had opposite signs. For DDT, this gradient was shown to be sensitive to input parameters, in particular solubility and photochemical degradation rate. The usage of a higher degradation rate closer to the upper limit of $k_{OH} = 1.5 \cdot 10^{-12} cm^3 molec^{-1} s^{-1}$ derived by Lammel et al (2009) has the potential of reducing the model error, and hence should be used in future simulation. Solubility of DDT is known only with high uncertainty. This uncertainty needs to be reduced in order to obtain reliable model results. The usage of the much lower solubility, then used in this study, suggested in Shen and Wania (2005) should be considered to be used in future simulations, as it potentially brings model results closer to observations. A lacking degradation process in the particle phase in the atmosphere was suggested as a reason for errornous model results. To enhance knowledge about a possible degradation of DDT in the particle phase laboratory studies and field measurements are needed.

Future work on MPI-MCTM should strive to improve parameterisations implemented in the model, because of two reasons: a) as a consequence of what was learned from the model evaluation with DDT, and b) more generally to obtain a better representation of key processes, which at least partly are insufficiently validated.

The comparison of predicted and observed DDT and DDE concentrations revealed that volatilisation from soil is one process that needs to be evaluated, as it potentially introduced errors into the model. In the current parameterisation the volatisation rate is empirically based upon establishment of equilibrium in the soil multiphase system [Smit et al (1997)]. A validation of the parameterisation for SOCs, e.g. DDT, and a comparison with other parameterisations using a MBM could provide a way to enhance understanding of the MPI-MCTM model results.

With regard to air-sea exchange, the current parameterisation is based on the two film model [Whitman (1923)], which describes gas exchange by molecular diffusion through a boundary layer, which is empirically related to wind speed. In stronger wind conditions (> 10 m/s) marine aerosol formation by bursting air bubbles can become important for gas exchange. This might be an important exchange process for less volatile pollutants, whose air-sea exchange is underestimated by considering only molecular diffusion. In particular, for PFOA it was suggested, that aerosol enrichment may contribute significantly to long-range transport, as these aerosols are transported [McMurdo et al (2008)]. Furthermore, the air-sea exchange parameterisation consideres only wind as the driving parameter for generation of turbulence.

Ho et al (2004) showed for SF_6 that in low wind conditions short intense rain events accelerate gas exchange in the marine environment. Extending air-sea exchange by including this process is also of interest in the context of climate change, as climate projections suggest that the intensity of rain events will increase [Roeckner et al (2006)].

The Earth system model ECHAM5-HAM/MPIOM-HAMOCC, upon which MPI-MCTM is based, is capable of reproducing the observed climate trend of the 20th century and was used to perform climate projections for the Fourth Assessment Report of the Intergovernmental Panel on Climate Change [Roeckner et al (2006), Intergovernmental Panel on Climate Change (2007)]. Principally this provides the prerequisite to use MPI-MCTM to answer questions of pollutant fate and distribution in a changing climate. Scenario-based simulations under future climate are justified for substances with similar properties as the ones studied in this study, i.e. DDT and PFOA , and should be based on historic multi-year simulations for contaminants with distinctly different properties, such as, for example, organochlorine pesticides of higher volatility than DDT or water-soluble organics considerably less polar than PFOA in order to evaluate MPI-MCTM's ability to reproduce their historical distributions. Climate predictions suggest that precipitation will increase in the humid climate zones (tropics, mid-high latitudes) and will decrease in arid climate zones [Roeckner et al (2006)]. An intensification of precipitation would presumably lead to an enhanced wash-out close to the sources for pollutants emitted in the tropics, thus leading to a reduced long-range transport of these pollutants. Thermal expansion and melting of Arctic inland ice is predicted to cause a sea level rise, and ground air temperatures are predicted to increase, especially in high northern latitudes. These changes, in turn, could trigger contaminant re-mobilisation from soils due to wash-out by sea level rise and enhanced volatilisation e.g. in northern Siberian soils. Re-mobilisation is also likely to occur due to melting of glaciers. Melting glaciers in Antarctica may re-mobilise persistent organic pollutants [Geisz et al (2008)]. Such a study would necessitate an implementation of inland ice as a compartment for chemicals cycling, as suggested by Guglielmo (2008). Currently the model only represents ice in the form of sea-ice.

Appendix A
Impact of the horizontal resolution on the representation of continental shelves

Minimal and maximal shelf areas in GR15 are derived from minimal and maximal distances of the grid cell corners.

Table A.1: Minimal and maximal areas covered by continental shelf (CS) in GR15.

Box	Shelf zone	area_{min} [km^2]	area_{max} [km^2]	Δarea [km^2]
I	Patagonian CS	927 877	1 220 680	292 803
II	Amazon CS	294 308	314 677	20 369
III	Gulf of Mexico	434 246	1 023 237	588 991
IV	Gulf of St.Lawrence	556 080	1 202 733	646 653
V	Hudson Bay	79 026	1 856 218	1 777 192
VI	Gulf of Alaska CS	317 463	1 421 607	1 104 144
VII	European CS	212 166	1 638 742	1 426 576
VIII	Mediterranean Sea	108 830	405 673	296 843
IX	Central Africa	108 890	204 281	95 391
X	East China Sea CS	882 687	986 200	103 513
XI	Sunda Shelf	2 994 864	3 567 390	572 526
XII	Australia	1 363 095	1 801 815	438 720
XIII	Sea of Okhotsk	441 967	579 088	137 121
XIV	CS of Russia	883 573	4 219 712	3 336 139

Appendix B
Sensitivity of volatilisation of DDT from the ocean to climate parameters

The partial correlation coefficient between two variables 1, 2 with a third variable 3, whose influence on 1 and 2 shall be excluded from the correlation is derived from [Sachs (1968)]:

$$R_{12.3} = \frac{R_{12} - R_{13}R_{23}}{\sqrt{(1 - R_{13}^2)(1 - R_{23}^2)}} \tag{B.1}$$

where R_{ij}; $i, j \in \{1, 2, 3\}$ denotes a linear correlation coefficient. The partial correlation coefficients for four possibly interacting variables is derived from

$$R_{12.34} = \frac{R_{12.4} - R_{13.4}R_{23.4}}{\sqrt{(1 - R_{13.4}^2)(1 - R_{23.4}^2)}} \tag{B.2}$$

where R_{ij} $i, j \in \{1, 2, 3\}$ denote linear correlation coefficients, and $R_{ij.k}$ $i, j, k \in \{1, 2, 3\}$ partial correlation coefficients derived from Equation B.1. The comparison of the linear correlation and the partial correlation can be used to examine the relationship between two variables, when additional variables are assumed to have a controlling influence [Ellet and Ericson (1986)]. Possible relationships for the correlation between two variables with a controlling variable are illustrated in Figure B.1. If both, linear and partial correlation coefficient are equal, the controlling variable has no effect. If the linear correlation is different from zero, but the partial correlation is zero, the linear correlation is spurious. It that case R_{12} can be explained by the correlations R_{13} and R_{23}. In a case of the linear correlation being positive and larger than the partial correlation, the linear correlation is partially explained by the correlations R_{13} and R_{23}, and, hence, overestimated. If the absolute value of the partial correlation is larger then the one of the linear correlation, it indicates that the linear correlation is suppressed due to the correlations R_{13} and R_{23}. If in that case R_{12} is zero, it is a spurious suppression, i.e. 3 is a consequent variable. It should be noted that, there is no positive proof for a causal relationship between variables from correlation analysis.

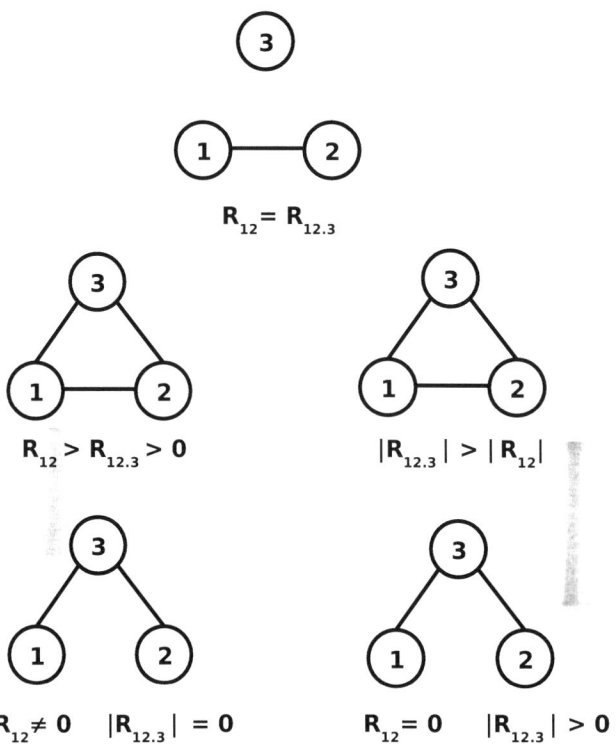

Fig. B.1: Relations between the linear and partial correlation coefficients for two variables and a controlling variable. Lines denote causal relations between the variables.

Figures

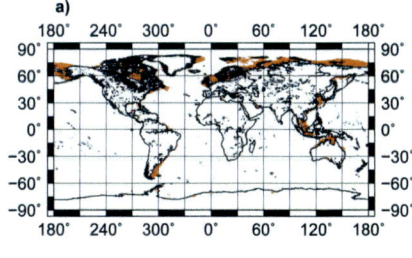 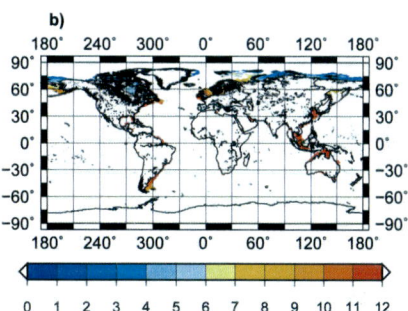

Fig. 2.1: a) Ocean waters with water depth lower than 250 m b) Total number of month per year in which MERIS data assimilation takes place.

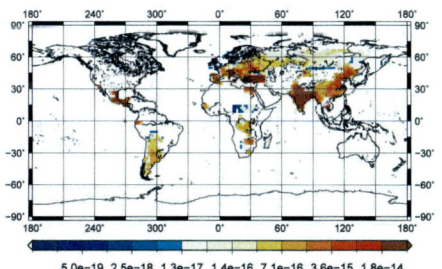

Fig. 2.4: DDT applications in 1980 [kg m^{-2} s^{-1}]. Applications were prepared by Semeena and Lammel (2003) from data reported to the UN Food and Agriculture Organisation (FAO).

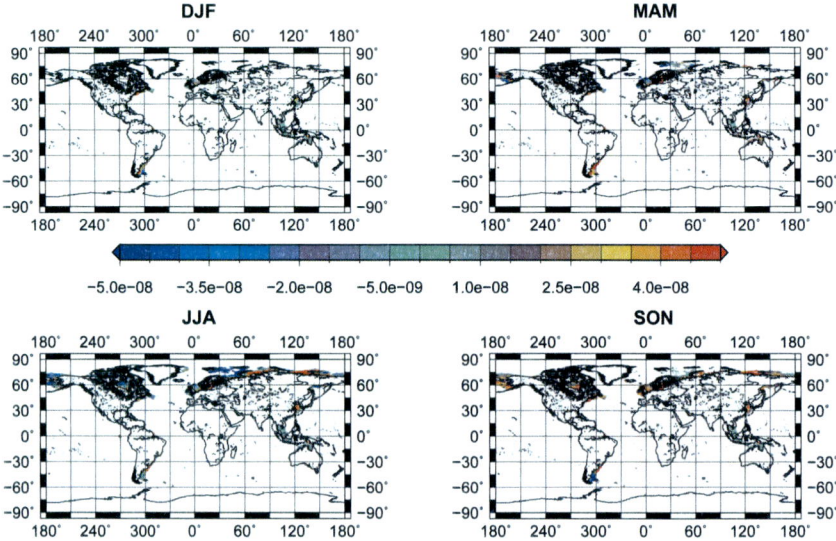

Fig. 2.6: Difference of the monthly mean phytoplankton concentration derived from MERIS data and the monthly mean calculated by HAMOCC [kmolP/m^3], denoted as 'a' in Equation 2.1.

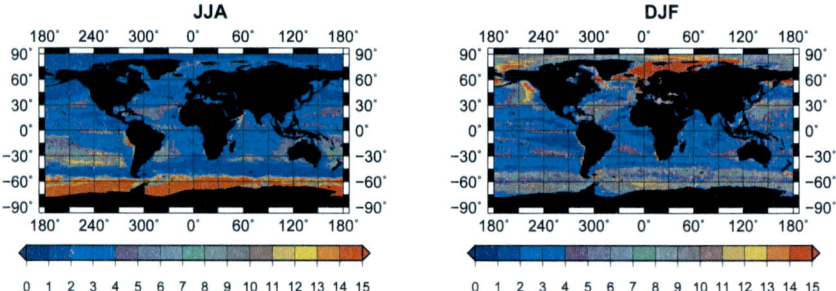

Fig. 2.9: Seasonal mean sinking velocity in the euphotic zone [m/d] in the season from June to August (JJA), and in the season from December to February (DJF).

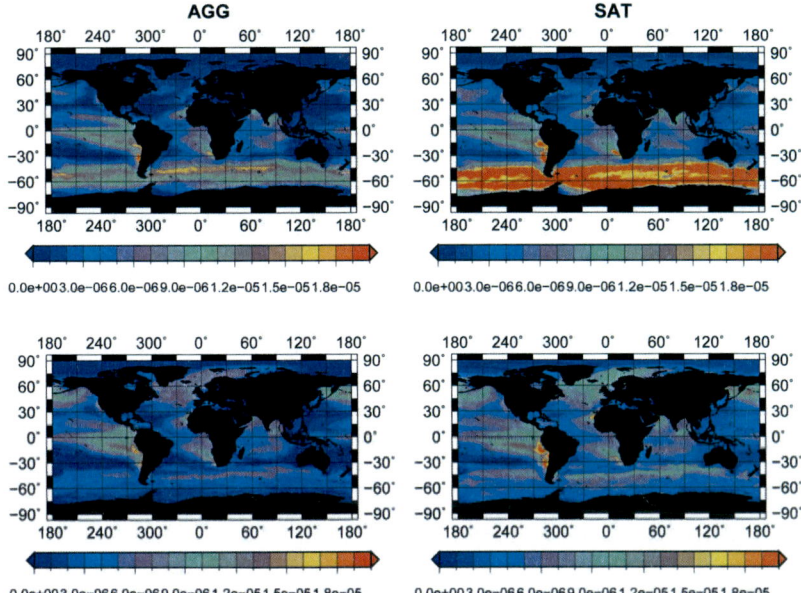

Fig. 2.10: Mean colloidal organic matter (COM - zooplankton+phytoplankton+dissolved organic carbon) burden [kmol P/m^2], integrated over the uppermost 90 m (euphotic zone).

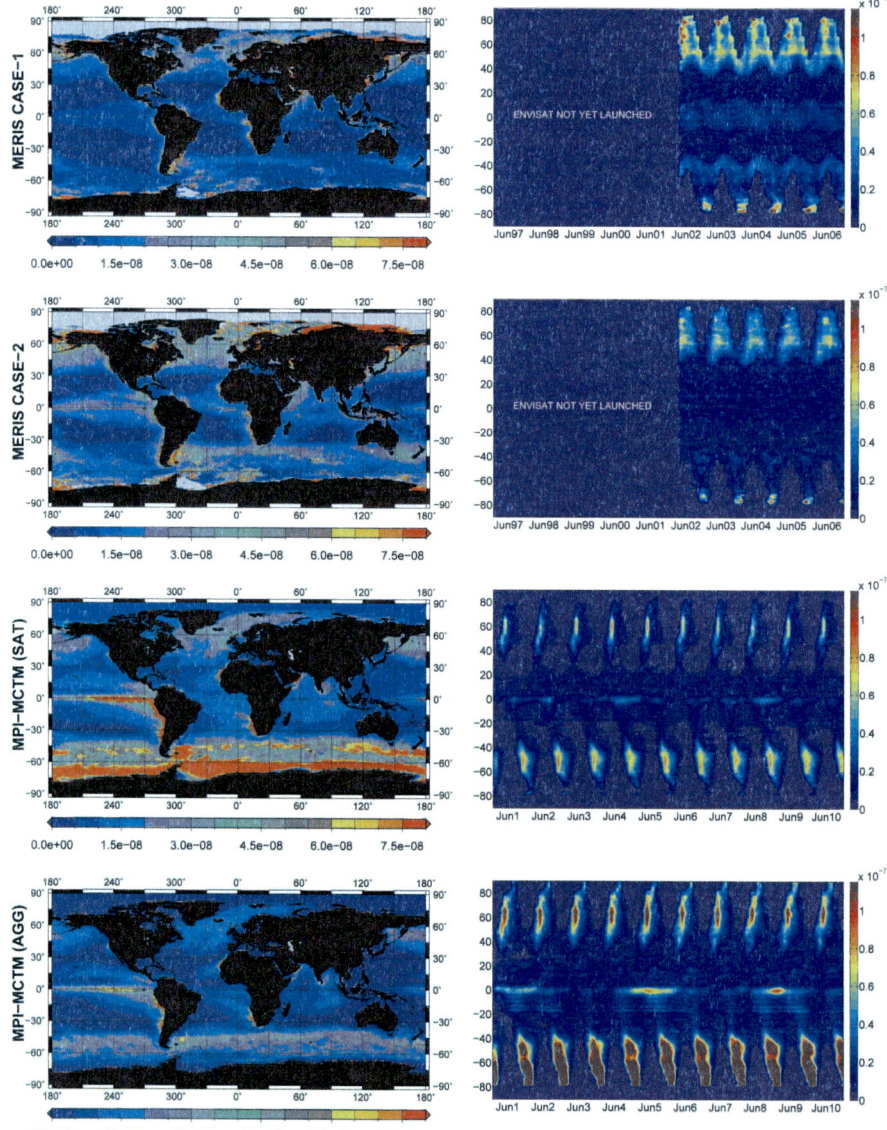

Fig. 2.11: Surface phytoplankton maps [kmol/m³] of overall temporal mean (left) and twodimensional time series (Hovmöller diagrams) showing the zonal mean seasonal variability.

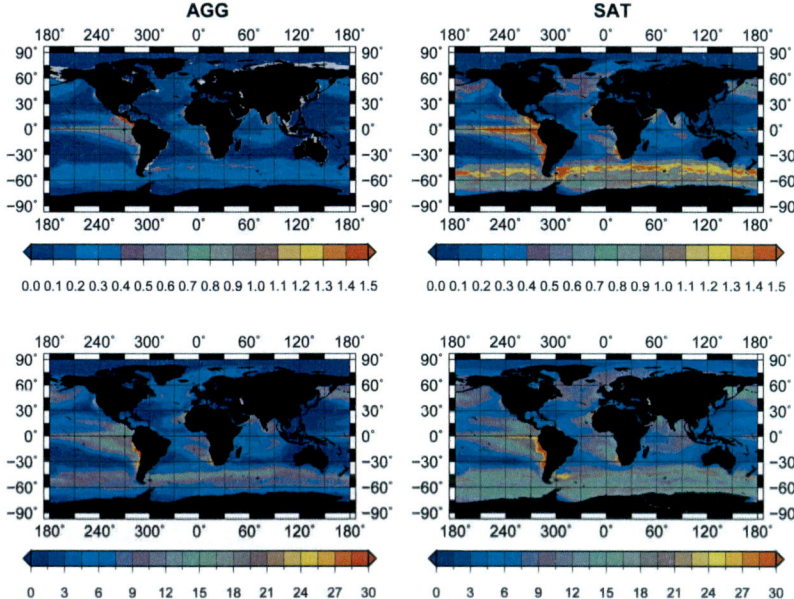

Fig. 2.13: Mean fraction of DDT bound on detritus (upper panels) and colloidal organic matter (lower panels) in the euphotic zone [%].

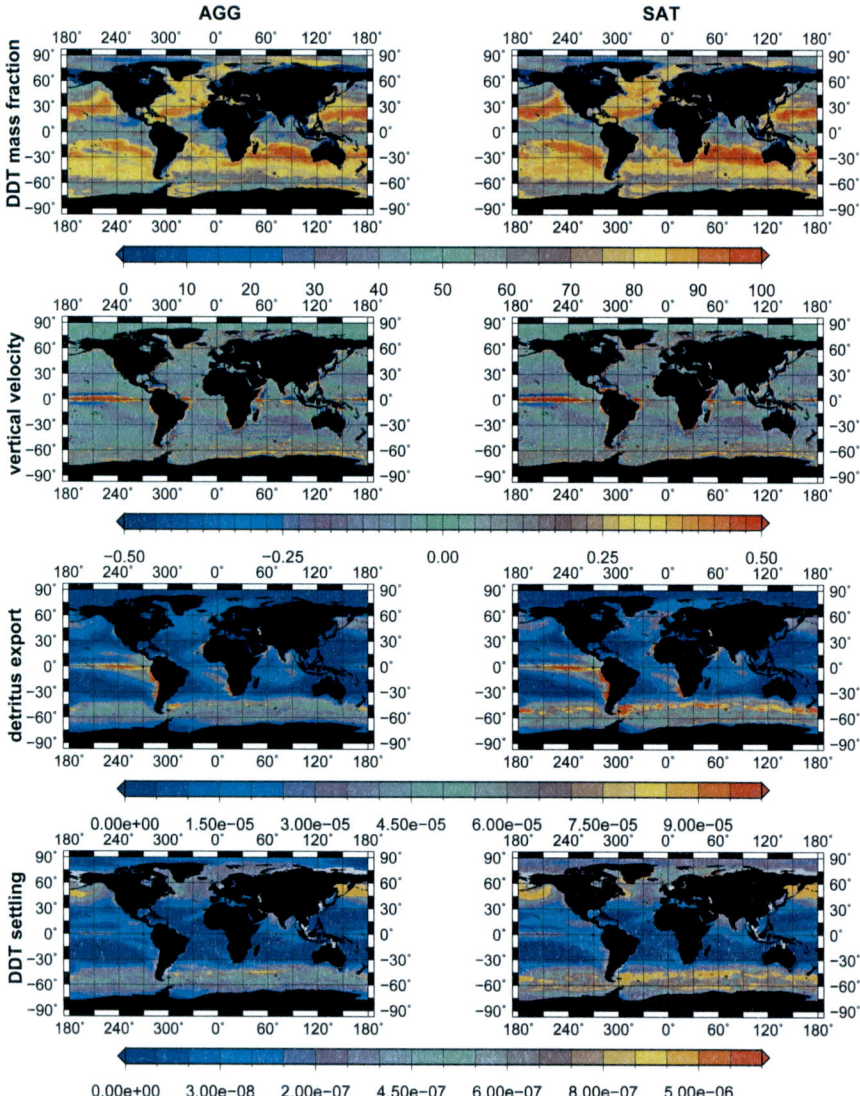

Fig. 2.14: DDT mass fraction below euphotic zone [%], mean vertical velocity in the euphotic zone [m/d], integrated detritus export out of the euphotic zone [kmol(C)/m²], and DDT downward flow (settling) [kg(DDT)/m²].

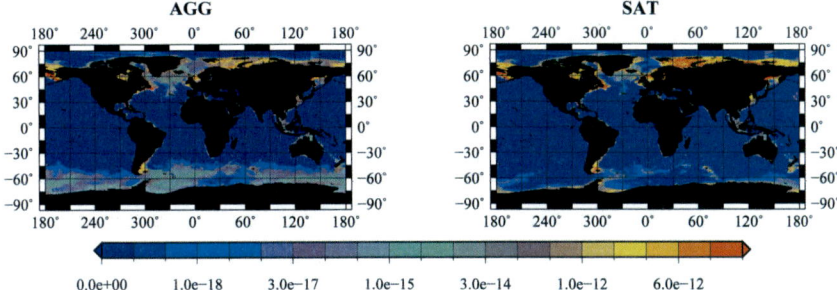

Fig. 2.15: Degradation in sediment [kg/m^2], accumulated over the simulation (10 years).

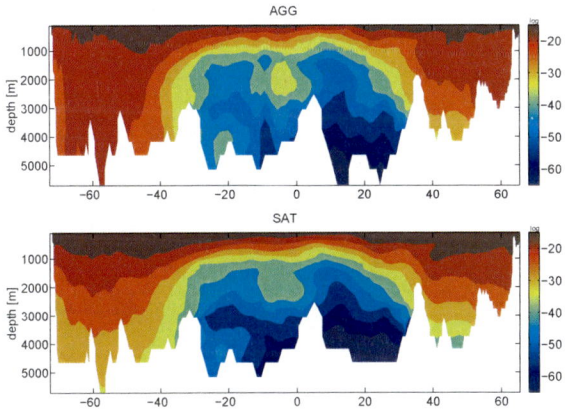

Fig. 2.16: Cross section of DDT concentration in the particulate phase [ng/L]. Transsect through the Atlantic Ocean.

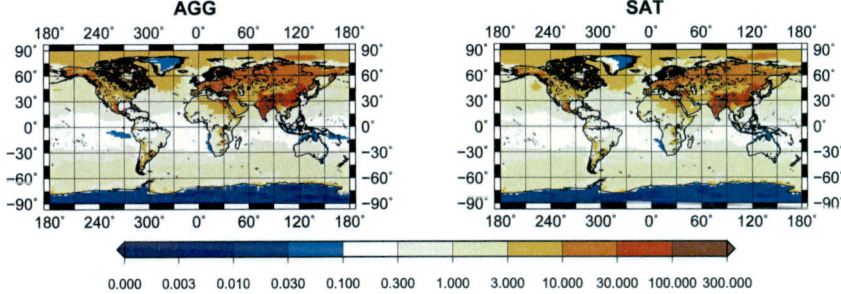

Fig. 2.19: Total environmental burden [kg/m^2] in December of the 10th year of the simulation.

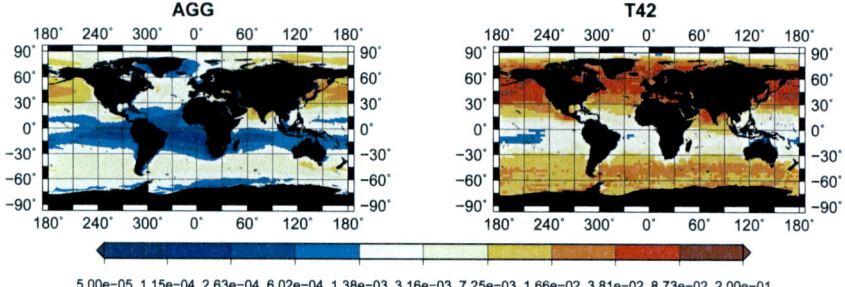

Fig. 2.20: Oceanic burden gradients [%]. Left panel shows results from the AGG, right panel shows results from Guglielmo (2008).

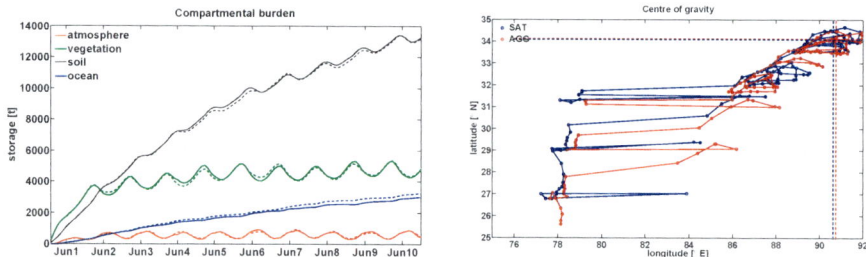

Fig. 2.21: Compartmental burden [t] (left panel), solid lines: model experiment with aggregation of marine snow (AGG), dashed lines: experiment with satellite assimilation (SAT). Migration of the centre of gravity of the total environmental burden (right panel). Dashed lines show the location of the COG at the end of the simulation. The COG of the SAT experiment is shown in blue, the COG of the AGG experiment in red. Circles represent monthly mean COGs.

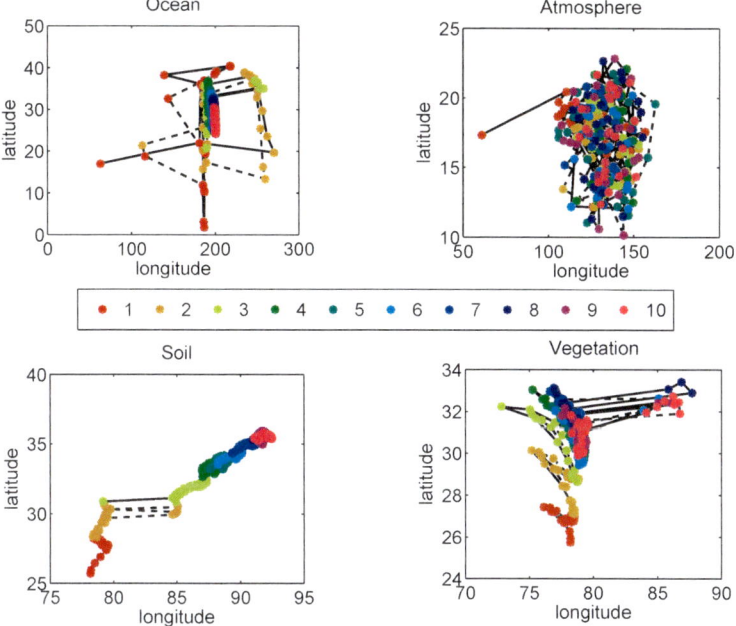

Fig. 2.22: Migration of the center of gravity of the DDT burdens in the individual compartments. Each asterisk represents a position derived from monthly mean data. Solid lines: model experiment with aggregation of marine snow (AGG), dashed lines: experiment with satellite assimilation (SAT), colour coded: years.

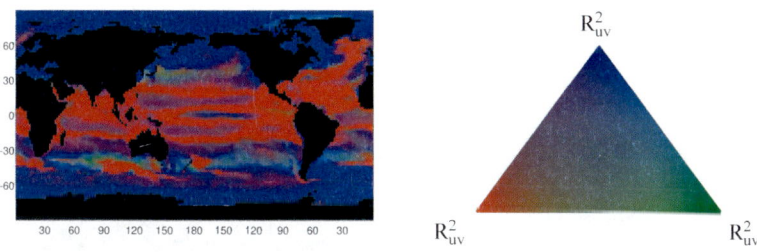

Fig. 2.26: RGB composite of normalised R^2 of the simple linear correlation coefficient, red associated with R^2_{uv}, blue with R^2_{tv}, green with R^2_{cv}.

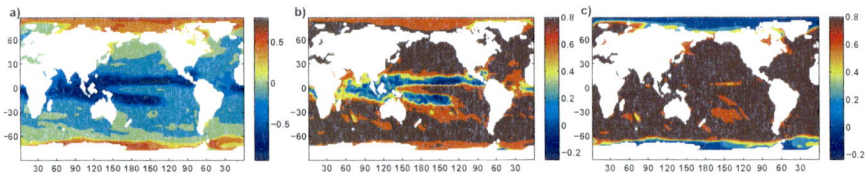

Fig. 2.27: Absolute differences of the coefficients of determination of partial correlations a) $R^2_{tv.uc} - R^2_{uv.tc}$ b) $R^2_{tv.uc} - R^2_{cv.ut}$ c) $R^2_{uv.tc} - R^2_{cv.ut}$.

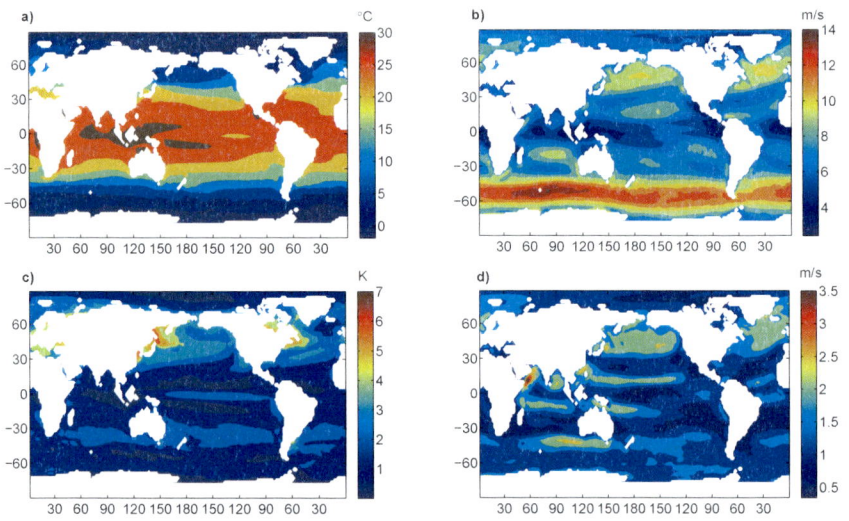

Fig. 2.28: 10-year means (a, b) and standard deviations (c, d) of SST (a, c) and 10 m wind speed (b, d).

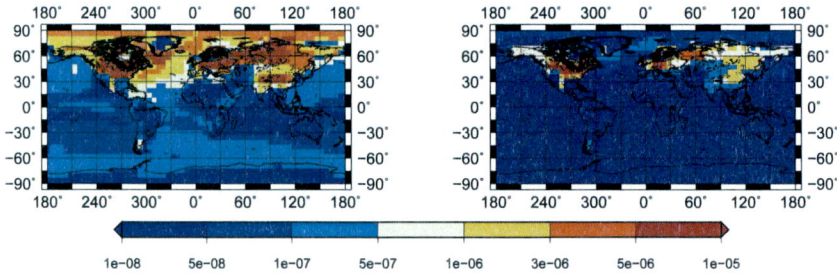

Fig. 3.2: Total environmental burden [kg/m²], December 1990 of DDT (left) and DDE (right).

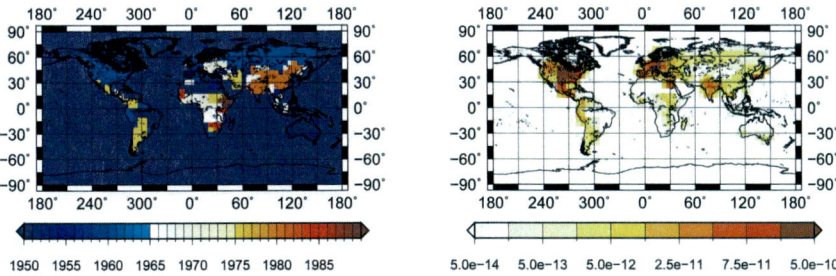

Fig. 3.3: Year of maximum DDT application [kg/m^2] (left) and accumulated for 1950-1990 (right).

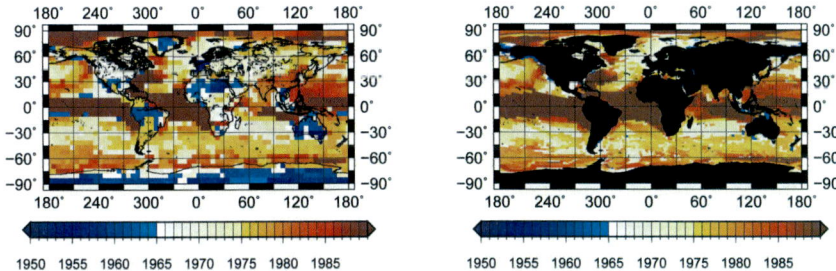

Fig. 3.4: Year in which the maximal DDT burden is reached, left: environmental burden , right ocean burden.

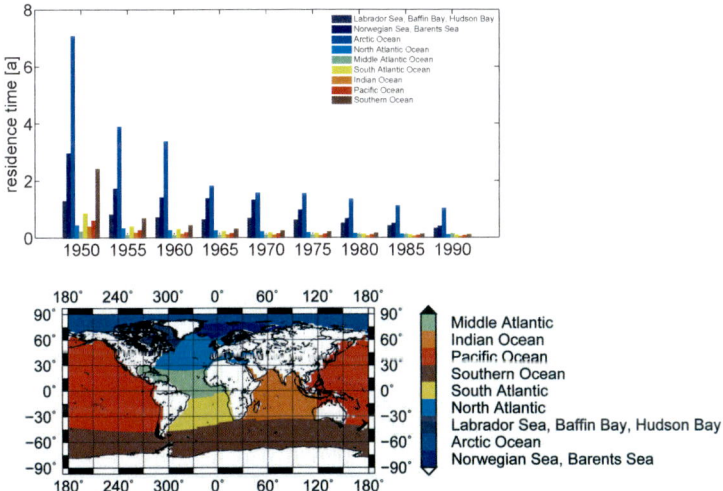

Fig. 3.6: Residence time [a] of DDT in various ocean regions.

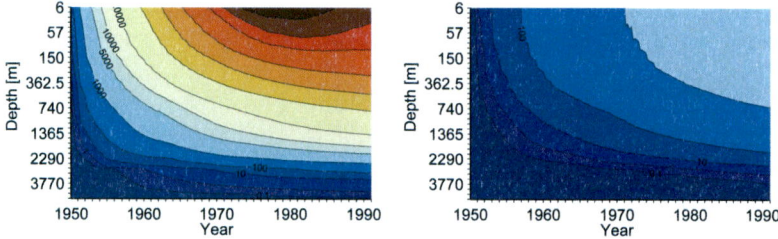

Fig. 3.7: Timeseries of the vertically integrated global DDT [t]. Integration starts from sea ground and goes up to the depths given in the axis label.

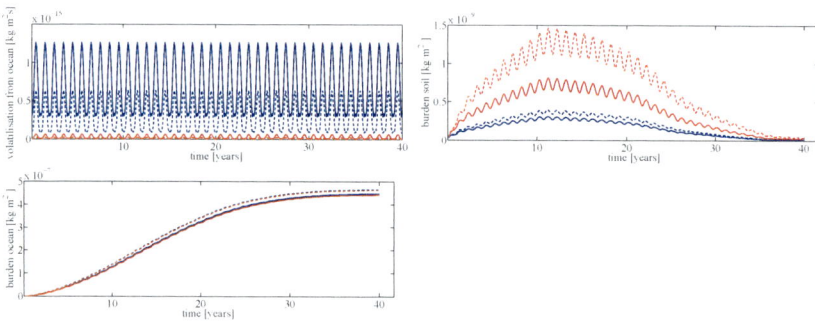

Fig. 3.10: DDT volatilisation flux from the ocean, soil and ocean burdens from MPI-MBM run assuming low (0.0034 mg/l) (blue) or high (0.1 mg/l) (red) water solubility under two climate scenarios represented by the surface temperatures T_{mean}=287.7 K (dashed) and T_{mean}=297.7 K (solid).

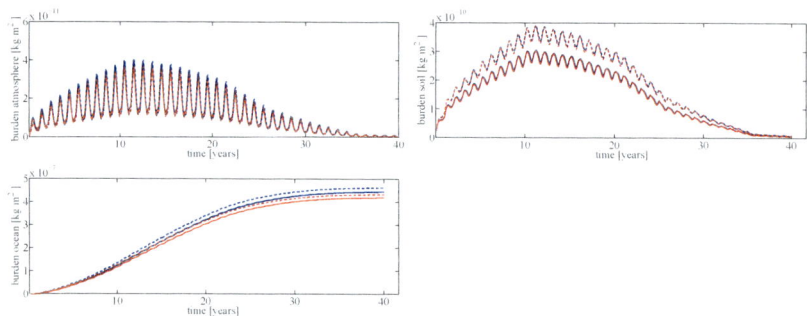

Fig. 3.11: DDT stored in atmosphere, soil and ocean as predicted by MPI-MBM with varying photochemical degradation rates $k_{OH} = 1.0 \cdot 10^{-13} \mathrm{cm^3 molec^{-1} s^{-1}}$ (blue) and $k_{OH} = 1.5 \cdot 10^{-12} \mathrm{cm^3 molec^{-1} s^{-1}}$ (red) under two climate scenarios represented by different surface temperatures, T_{mean}=287.7 K (dashed) and T_{mean}=297.7 K (solid).

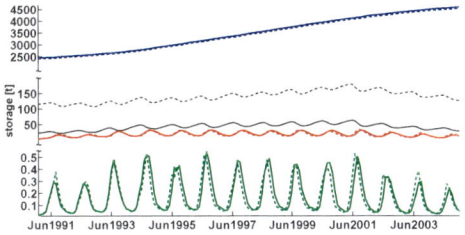

Fig. 3.13: Global burdens of PFOA (KOC115 (solid line),KOC11500 (dashed line)) in atmosphere (red), vegetation (green), soil (black) and ocean (dark blue). Evolution of the monthly mean in [t].

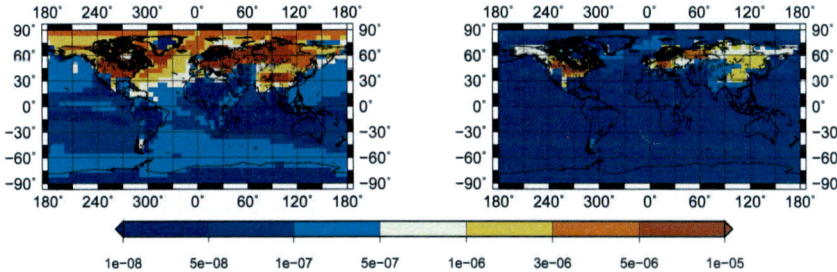

Fig. 3.14: Oceanic burden [kg/m2], December 2004. Left KOC115, right KOC11500.

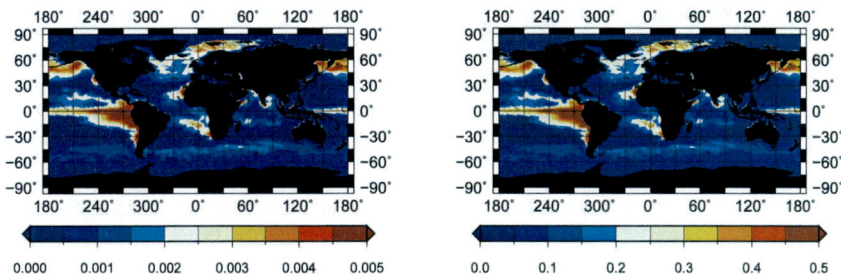

Fig. 3.16: Fraction bound to organic matter at the surface [%], June 2004. Left KOC115, right KOC11500.

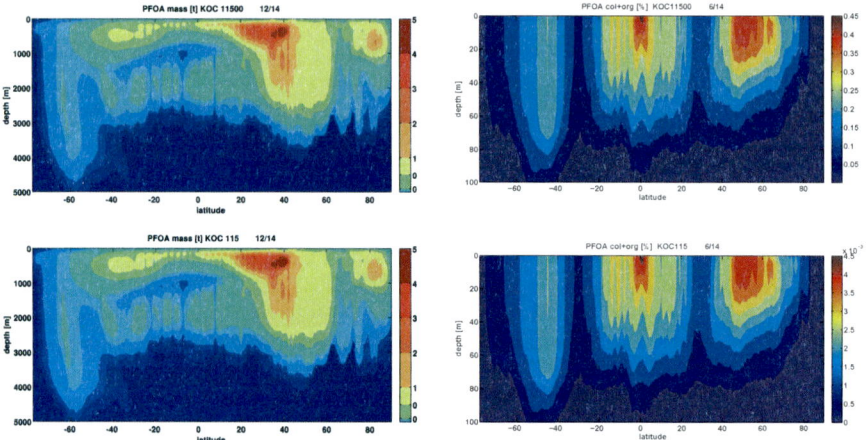

Fig. 3.17: Vertical and latitudinal distribution of PFOA mass in the global oceans [t] (left),12/2004, fraction of mass bound to organic matter [%] (right), 06/2004.Upper panels KOC115, lower panels KOC11500.

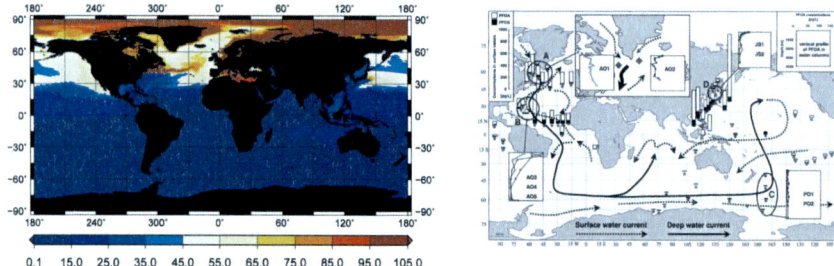

Fig. 3.18: Ocean surface layer concentrations of PFOA [pg/L], left: annual mean 2004, right: merged observations from 2002 to 2006 Yamashita et al (2008).

References

ACRI-ST, Laboratoire Océanologique de Villefranche-sur-mer (LOV), University of Plymouth, UK (UoP), Norwegian Institute for Water Research (NIVA), Brockmann-Consult (BC), Deutsches Zentrum für Luft- und Raumfahrt (DLR), Institute for Computational and Earth Systems Science (ICESS) (2007) GlobColour: An EO based service supporting global ocean carbon cycle research. Product User Guide. GC-UM-ACR-PUG-01, http://www.globcolour.info/products_description.html

Addison RF, Willis DE (1978) The metabolism by rainbow trout (salvo garidnerri) of p,p-[^{14}C]DDT and some of its possible degradation products labeled with ^{14}C. Toxicology and Applied Pharmacology 43:303–315

Aisalbie JM, Richards NK, Bould HL (1997) Microbial degradtion of DDT and its residues - a review. New Zeal JAgric Res 40:269–282

Alldredge AL, Gotschalk CC (1989) Direct observations of the mass flocculation of diatom blooms: characteristics, settling velocities and formation of diatom aggregates. Deep-Sea Reserach 36:1959–171

Arakawa A, Lamb VR (1977) Computational design of the basic dynamical processes of the UCLA general circulation model. Methods Comput Phys 17:173–265

Arctic Monitoring and Assessment Programme (AMAP) (2004) AMAP assessment 2002: persistent organic pollutants in the Arctic. AMAP, P.O. Box 8100 Dep, N-0032 Oslo, Norway, www.amap.no

Armitage J, Cousins IT, Buck RC, Prevedouros K, Russell MH, MacLeod M, Korzeniowski SH (2006) Modeling global-scale fate and transport of perfluorooctanoate emitted from direct sources. Environ Sci Technol 40:6969–6975

Arp HPH, Niederer C, Goss KU (2006) Predicting the partitioning behaviour of various higly fluorinated compounds. Environ Sci Technol 40:7298–7304

Asper VL (1987) Measuring the flux and sinking speed of marine snow aggregates. Deep-Sea Research 34.1–17

Atkinson R, Guicherit R, Hites RA, Palm WU, Seiber JN, de Voogt P (1999) Transformations of pesticides in the atmosphere: a state of the art. Water, Air and Soil Pollution 115:219–243

Azetsu-Scott K, Johnson BD (1992) Measuring physical charcteristics of particles: a new method of simultaneous measurements for size , settling velocity and density of constituent matter. Deep-Sea Research 39:1057–1066

Biggar JW, Dutt GR, Riggs RL (1967) Predicting and measuring the solubility of p,p'-DDT in water. Bulletin of Environmental Contamination and Toxicology 2:90–100

Boethling RS, Mackay D (2000) Handbook of property estimation methods for chemicals: environmental and health science. Lewis Publishers, Boca Raton London New York Washington D.C.

Boulanger B, Peck AMea (2005) Mass budget of perfluorooctane surfactants in lake ontario. Environ Sci Technol 39:74–79

Brace NO (1962) Long chain alkanoic and alenoic acids with perfluoroalkyl terminal segments. J Org Chem 40:4491–4493

Carder KI, Steward RG, Betzer BR (1982) In situ holographic of the size and settling rates of oceanic particles. Journal of Geophysical Research 87:5681–5685

Cherry RD, Higgo JJW, Fowler SW (1978) Zooplankton fecal pellets and element residence times in the ocean . Nature 274:246–248

Conder JM, Hoke RA, de Wolf W, Russel MH, Buck RC (2008) Are PFCAs bioaccumulative? A critical review and comparison with regulatory criteria and persistent lipophilic compounds. Environ Sci Technol 42:995–1003

Cramer J (1973) Model of the circulation of DDT on Earth. Atmospheric Environment 7:241–256

Cziudaj G (2005) Die Rolle der Ozeane bei der globalen Verteilung von organischen Schadstoffen: PCB als Modellsubstanzen. PhD thesis, Christian Albrechts University Kiel

Dachs J, Bayona JM, Fowler SW, Miquel J, Albaigés J (1996) Vertical fluxes of polycyclic aromatic hydrocarbons and organochlorine compounds in the western Alboran Sea (southwestern Mediterranean). Marine Chemistry 52:75–86

Dachs J, Bayona JM, Albaigs J (1997) Spatial distribution, vertical profiles and budget of organochlorine compounds in Western Mediterranean seawater. Marine Chemistry 57(3-4):313 – 324

Dachs J, Eisenreich SJ, Baker JE, Ko FC, Jeremiason JD (1999) Coupling of phytoplankton uptake and airwater exchange of persistent organic pollutants. Environ Sci Technol 33:3653–3660

Dachs J, Lohmann R, Ockenden WA, Mejanelle L, Eisenreich S, Jones KC (2002) Oceanic biogeochemical controls on global dynamics of persistent organic pollutants. Environ Sci Technol 36:4229–4237

De Wolf, W, De Bruijn, J H M, Seinen W, Hermens JLM (1992) Influence of biotransformation on the relationship between bioconcentration factors and octanol water partition-coefficients. Environ Sci Technol 26(6):1197–1201

Dimond JB, Owen RB (1996) Long-term residue of DDT compounds in forest soils in Maine. Environmental Pollution 92:227–230

Doerffer R, Schiller H (2000) Neural network or retrieval of concentrations of water constituents with the possibility of detecting exceptional out of scope spectra. Proceedings of IGARSS pp 714–717

Doerffer R, Schiller H (2007) The MERIS Case 2 water algorithm. International Journal of Remote Sensing 28:517–535

Eggleton J, Thomas KV (2004) A review of factors affecting the release and bioavailability of contaminants during sediment disturbance events. Environement International 30:973–980

Ellet FS, Ericson DP (1986) Correlation, partial correlation and causation. Synthese 67:157–173

Ellis DA, Mabury SA, Martin JW, Muir DCG (2001) Thermolysis of fluoropolymers as a potential source of halogenated organic acids in the environment. Nature 412:321–324

Ellis DA, Martin JW, De Silva AO, Mabury SA, Hurley MD, Sulbaek Anderson MP, Wallinton TJ (2004) Degradation of fluorotelomer alcohols: a likely atmospheric source of perfluorinated carboxylic acids. Environ Sci Technol 38:3316–3321

Ernst W, Goerke H (1974) Anreicherung, Verteilung, Umwandlung und Ausscheidung von DDT-^{14}C bei Solea solea (Pisces : Soleidae). Marine Biology 24:287–304

Falandysz J, Taniyasu S, Yamashita N, Rostkowski P, Zalewski K, Kannan K (2007) Perfluorinated compounds in some terrestrial and aquatic wildlife species from Poland. J Environ Sci Health 42A:715–719

Finizio A, Mackay D, Bidleman T, Harner T (2007) Octanol-air partition coefficient as a predictor of partitioning of semi-volatile organic chemicals to aerosols. Atmospheric Environment 31:2289–2296

Fowler SW, Knauer GA (1986) Role of large particles in the transport of elements and organic compounds through the oceanic water column. Prog Oceanog 16:147–194

Ganzeveld L, Lelieveld J (1995) Dry deposition parameterization in a chemistry general-circulation model and its influence on the distribution of reactive trace gases. J Geophys Res 100(D10):20,999–21,012

Geisz HN, Dickhut RM, Cochran MA, Fraser WR, Ducklow HW (2008) Melting glaciers: A probable source of DDT to the Antarctic marine ecosystem . Environ Sci Technol 42:3958–3962

Gent PR, Willebrand J, McDougall T, McWilliams JC (1995) Parameterizing eddy-induced tracer transports in ocean circulation models. J Phys Oceanogr 106 (C2):2693–2712

Giesy JP, Kannan K (2002) Perfluorochemical surfactants in the environment. Environ Sci Technol 36:146A–152A

Gonzalez RC, Woods RE (1993) Digital image processing. Addison-Wesley, Reading, USA

Gordon HR, Morel AY (1983) Remote assessment of ocean colour for interpretation of satellite visible imagery: A review . Springer, New York, USA

Gramatica P, Papa E (2007) Screening and ranking of POPs for global half-life: QSAR approaches for prioritization based on molecular structure. Environ Sci Technol 41(8):2833–2839

Griffies SM (1998) The Gent-McWilliams skew flux. J Phys Oceanogr 28:831–841

Guglielmo F (2008) Global cycling of semivolatile organic compounds in the marine and total environment - a study using a comprehensive model. PhD thesis, MPI Reports on Earth System Science No.54

Gustafsson O, Gschwend PM (1997) Aquatic colloids: Concepts, definitions, and current challenges. Limnol Oceanogr 42:519–528

Gustafsson O, Gschwend P, Buesseler K (1997) Settling removal rates of PCBs into the Northwestern Atlantic derived from U-238-Th-234 disequilibria. Environ Sci Technol 31(12):3544–3550

Haque R, Freed VH (eds) (1975) Environmental dynamics of pesticides. Plenum Press, New York

Hecht M, Hasumi H (eds) (2008) Ocean modeling in an eddying regime. American Geophysical Union, Geophysical Monograph Series Vol.177, Washington DC, USA

Hellström A, Kylin H, Strachan WMJ, Jensen S (2004) Distribution of some organochlorine compounds in pine needles from Central and Northern Europe. Environ Poll 128:29–48

Hibler WD (1979) A dynamic thermodynamic sea ice model. J Phys Oceanogr 9:815–844

Higgins CP, Luthy RG (2006) Sorption of perfluorinated surfactants on sediments. Enivorn Sci Technol 40:7251–7256

Ho DT, Zappa CJ, McGillis W, Bliven LF, Ward B, Dacey JWH, Schlosser P, Hendricks MB (2004) Influence of rain on air-sea gas exchange: Lessons from a model ocean. J Geophysical Research C08S18

Hornsby AG, Wauchope DR, Herner AE (1996) Pesticide properties in the environment. Springer, New York, USA

Horowitz L, Walters S, Mauzerall D, Emmons L, Rasch P, Granier C, Tie X, Lamarque J, Schultz M, Tyndall G, Orlando J, Brasseur G (2003) A global simulation of tropospheric ozone and related tracers: Description and evaluation of MOZART, version 2. Journal of Geophysical Research 108(D24), DOI 10.1029/2002JD002853

Hurley MD, Andersen MPS, Wallington TJ, Ellis DA, Martin JW, Mabury SA (2004) Atmospheric chemistry of perfluorinated carboxylic acids: Reaction with OH radicals and atmospheric lifetimes. J Phys Chem A 108(4):615–620, DOI 10.1021/jp036343b

Intergovernmental Panel on Climate Change (2007) Climate Change 2007: Synthesis Report, Contribution of Working Groups I, II and III to the Fourth Assessment Report of the Intergovernmental Panel on Climate Change. http://www.ipcc.ch/ipccreports/ar4-syr.htm

IPCS (1979) DDT and its derivatives. Environemntal Health Criteria 9: http://www.inchem.org/documents/ehc/ehc/ehc009.htm

Iwata H, Tanabe S, Sakai N, Tatsukawa R (1993) Distribution of persistent organochlorines in the oceanic air and surface seawater and the role of ocean on their global transport and fate. Environ Sci Technol 27:1080–1098

Jeffrey SW, Mantoura RFC, Wright SW (1997) Phytoplankton Pigments in Oceanography. UNESCO Publishing, Paris, France

Jensen S, Eriksson G, Kylin H (1992) Atmospheric pollution by persistent organic compounds: monitoring with pine needles. Chemosphere 24:229–245

Jurado E (2006) Modelling the ocean atmosphere exchanges of persistent organic pollutants (POPs). PhD thesis, Universitat Politècnica de Catalunya, Barcelona, Spain

Jurado E, Zaldvar JM, Marinov D, Dachs J (2007) Fate of persistent organic pollutants in the water column: Does turbulent mixing matter? Mar Poll Bull 57:441–451

Kissa E (2001) Fluorinated Surfactants and Repellents. Marcel Dekker Inc.

Klöpffer W, Schmidt E (2003) Comparative determination of the persistence of Semivolatile Organic Compunds (SOC) using SIMPLEBOX 2.0 and CHEMRANGE 1.0/2.1. Fresenius Environmental Bulletin 12:490–496

Kriest I (2002) Different parameterizations of marine snow in a 1d-model and their influence on representation of marine snow, nitrogen budget and sedimentation. Deep Sea Research I 49:2133–2162

Lammel G (2004) Effects of time-averaging climate parameters on predicted multicompartmental fate of pesticides and POPs. Environ Poll 128:291–302

Lammel G, Feichter J, Leip A (2001) Long-range transport and global distribution of semivolatile organic compounds: A case study on two modern agrochemicals. MPI Report No. 324

Lammel G, Klánová J, Kohoutek J, Prokeš R, Ries L, Stohl A (2009) Observation and origin of organochlorine compounds and polycyclic aromatic hydrocarbons in the free troposphere over Central Europe. Environ Poll in press

Lammel W Gand Klöpffer, Semeena VS, Schmidt E, Leip A (2007) Multicompartmental fate of persistent substances. Comparison of predictions from multi-media box models and a multi-compartment chemistry-atmospheric-transport model. Env Sci Pollut Res 14:153–165

Lau C, Butenhoff JL, Rogers JM (2004) The developmental toxicity of perfluoroalkyl acids and their derivatives. Toxicology and Applied Pharmacology 198:231–241

Lau C, Butenhoff JL, Rogers JM (2006) Effects of perfluorooctanoic acid exposure during pregnancy in the mouse. Toxicological Sciences 90:510–518

Leah RT, Johnson MS, L Connor L, Levene CF (1997) DDT group compounds in fish and shellfish from the Mersey Estuary and Liverpool Bay. Environmental Toxicology and Water Quality 12:223–229

Leip A, Lammel G (2004) Indicators for persistence and long-range transport potential as derived from multicompartment chemistrytransport modelling. Environ Poll 128:205–221

Lin SJ, Rood RB (1996) Multidimensional flux form semi-Lagrangian transport. Mon Wea Rev 124:2046–2068

Lohmann U, Roeckner E (1996) Design and performance of a new cloud microphysics scheme developed for the ECHAM4 general circulation model. Clim Dyn 12:557–572

Mackay D (1991) Multimedia Environmental Models: The Fugacity Approach. Lewis Publishers, Chelsea, MI, USA

Mackay D, Yeun ATK (1983) Mass transfer coefficients correlations for volatilisation of organic solutes from water. Environ Sci Technol 17:211–233

Maier-Reimer E, Kriest I, Segschneider J, Wetzel P (2005) The HAMburg Ocean Carbon Cycle Model HAMOCC5.1 - Technical Description Release 1.1 -. MPI Reports on Earth System Science No.14:1–57

Mann KH, Lazier JRN (2006) Dynamics of Marine Ecosystems. Blackwell Publishing

Marsland SJ, Haak H, Jungclaus JH, Latif M, Röske F (2003) The Max-Planck-Instistute global ocean-sea ice model with orthogonal curvelinear coordinates. Ocean Modelling 5:91–127

Martin S (2004) An introduction to ocean remote sensing. Cambridge University Press, Cambridge, UK

Maugh P (1973) An unrecognized source of polychlorinated biphenyls. Science 180:1527–1528

McCave IN (1984) Size spectra and aggregation of suspended particles in the deep ocean. Deep-Sea Res 31:329–352

McMurdo CJ, Ellis DA, Webster E, Butler J, Christensen RD, Reid LK (2008) Aerosol enrichment of the surfactant PFO and mediation of the water - Air transport of gaseous PFOA. Environ Sci Technol 42(11):3969–3974, DOI 10.1021/es7032026

Meijer SN, Halsall CJ, Harner T, Peters A, Ockenden WA, Johnston AE, Jones KC (2001) Organochlorine pesticide residues in archived UK soil. Environ Sci Technol 35:1989–1995

Morel A, Prieur L (1977) Analysis of variations in ocean colour. Limnology and Oceanography 22:709–722

Mork KA, Blindheim J (2000) Variations in the Atlantic inflow to the Nordic Seas, 1955-1996. Deep-Sea Research I 47:1035–1057

NOAA (1988) Etopo-5 bathymetry/topography data, data announcement 88mgg02. Tech. rep., National Oceanic and Atmospheric Administration, U.S. Department of Commerce, Boulder, USA

NOAA (2001) Etopo2 global 2' elevations cd-rom. Tech. rep., National Geophysical Data Center, NOAA/NGDC. USA, http://www.ngdc.noaa.gov/mgg/fliers/06mgg01.html

Nordeng TE (1994) Extended versions of the convective parameterization scheme at ECMWF and their impact on the mean and transient activity of the model in the tropics . Tech. rep., ECMWF, Reading, UK, technical memorandum edn.

O'Brien RD (1975) Nonenzymic effects of pesticides on membranes, pp 331–342. In: Haque and Freed (1975)

Oliaei F, Kriens D, Kessler K (2006) Investigation of perfluorochemical (pfc) contamination in minessota - phase one. Tech. rep., Report to Senate Environment Commitee

Olsen GW, Church TR, Miller JP, Burris JM, Hansen KJ, Lundberg JK, Armitage JB, Herron RB (2003) Perfluorooctanesulfonate and other fluorochemicals in the serum of American Red Cross adult blood donators. EnvironHealth Perspect 111:1892–1901

Olsen GW, Burris JM, Ehresman DJ, Froehlich JW, Seacat JW, Butenhoff JL, Zobel LR (2007) Half-life of serum elimination of perfluorooctanesulfonate, perfluorohexanesulfonate, and perfluorooctanote in retired fluorochemical production workers. Environ Health Perspect 155:1298–1305

Oschlies A (2008) Eddies and upper-ocean nutrient supply, pp 115–130. In: Hecht and Hasumi (2008)

Pacanowski RC, Philander SGH (1981) Parameterization of vertical mixing in numerical models of tropical oceans. J Phys Oceanogr 11:1443–1451

Patil KC, Matsumura F, Boush MG (1972) Metabolic transformation of DDT, Dieldrin, Aldrin, and Endrin by marine microorganisms. Environ Sci Technol 6:629–632

Patton GW, Hinckley DA, Walla MD, Bidleman TF (1989) Airborne organochlorines in the Canadian High Arctic. Tellus 41B:243–255

Peixoto JP, Oort AH (1992) Physics of Climate. Springer, New York, USA

Pontollilo J, Eganhouse RP (2001) The search for reliable aqueous solubility (S_w) and octanol-water partition coefficient (K_{ow}) data for hydrophobic organic compounds: DDT and DDE as a case study. US Geological Survey, Water Investigations Report 01-4201:1–51

Preisendorfer RW (1961) Application of radiative transfer theory to light measurements in the sea. IUGG Monographs 10:11–30

Prevedouros K, Cousins IT, Buck RC, Korzeniowski SH (2006) Sources, fate and transport of perfluorocarboxylates. Environ Sci Technol 40:32–44

Rast M (1999) The ESA Medium Resolution Imaging Spectrometer MERIS - a review of the instrument and its mission. International Journal of Remote Sensing 20:1679–1680

Redi MH (1982) Oceanic isopycnal mixing by coordinate rotation. J Phys Oceanogr 12:1154–1158

Ritter L, Solomon KR, Forget J, Stemeroff M, O'Leary C (1995) Persistent Organic Pollutants - An Assessment Report on: DDT-Aldrin-Dieldrin-Endrin-Chlordane Heptachlor-Hexachlorobenzene Mirex-Toxaphene Polychlorinated Biphenyls Dioxins and Furans. For: The International Programme on Chemical Safety (IPCS) within the framework of the Inter-Organization Programme for the Sound Management of Chemicals (IOMC) pp 1–43

Roeckner E, Bäuml G, Bonaventura L, Brokopf R, Esch M, Giorgetta S M Hagemann, Kirchner I, Kornblüh L, Manzini E, Rhodin A, Schlese U Uand Schulzweida, Tompkins A (2003) The atmospheric general circulation model ECHAM5 part 1: Model description. MPI Report No349, MPI for Meteorology, Hamburg

Roeckner E, Brasseur G, Giorgetta M, Jacob D, Jungclaus J, Reick C, Sillmann J (2006) Climate Prjections for the 21st century . MPI-M

Sachs L (1968) Statistische Auswertungsmethoden. Springer, Berlin

Schenker U, Scheringer M, Hungerbohler K (2008a) Investigating the global fate of DDT: Model evaluation and estimation of future trends. Environ Sci Technol 42:1178–1184

Schenker U, Scheringer M, MacLeod M, Martin JW, Cousins IT, Hungerbühler K (2008b) Contribution of volatile precursor substances to the flux of perfluorooctanoate to the Arctic. Environ Sci Technol 42(10):3710–3716

Scheringer M (1996) Persistence and spatial range as endpoints of an exposure-based assessment of organic chemicals. Environ Sci Technol 30:1652–1659

Scheringer M, Stroebe M, Wania F, Wegmann F, Hungbühler K (2004) The effect of export to the deep sea on the long-range transport potential of persistent organic pollutants. Environ Sci Poll Res 1:41–48

Schwarzenbach R, Gschwend P, Imboden D (2003) Environmental Organic Chemistry, Second Edition. John Wiley & Sons, Inc., New Jersey, USA

Semeena VS (2005) Long-range atmospheric transport and total environmental fate of persistent organic pollutants - a study using a general circulation model. PhD thesis, MPI Reports No. 15, MPI for Meteorology, Hamburg

Semeena VS, Lammel G (2003) Effects of various scenarios of entry of DDT and γ-HCH on the global environmental fate as predicted by a multicompartment chemistry-transport model. Fresenius Environmental Bulletin 12:925–939

Semeena VS, Lammel G (2005) Significance of the grasshopper effect on the atmospheric distribution of persistent organic substances. Geophysical Research Letters 32

Senthilkumar K, Ohi E, Sajwan K, Takasuga T, Kannan K (2007) Perfluorinated compounds in river water, river sediment, market fish, and wildlife samples from japan. Bull of Environ Contam Toxicol 79:427–431

Shen L, Wania F (2005) Compilation, evaluation and selection of physical-chemical property data for organochlorine pesticides. J Chem Eng Data 50:742–768

Skei J, Larsson P, Rosenberg R, Jonsson P, Olsson M, Broman D (2000) Eutrophication and contaminants in aquatic ecosystems. Ambio 29(4-5):184–194

Skoglund RS, Stange K, Swackhamer DL (1996) A kinetics model for predicting the accumulation of pcbs in phytoplankton. Environ Sci Technol 30(7):2113–2120, DOI 10.1021/es950206d, URL http://pubs.acs.org/doi/abs/10.1021/es950206d, http://pubs.acs.org/doi/pdf/10.1021/es950206d

Smit AAMFR, van den Berg F, Leistra M (1997) Estimation method for the volatilization from fallow soils. Environmental Planning Bureau series 2, Wageningen, Netherlands, DLO Winand Staring Centre

Smit AAMFR, van den Berg F, Leistra M (1998) Estimation method for the volatilization of pesticides from plants. Environmental Planning Bureau series 4, Wageningen, Netherlands, DLO Winand Staring Centre

Snedeker SM (2001) Pesticides and breast cancer risk: A review of DDT, DDE and Dieldrin. Environmental Health Perspectives 109:35–47

Spivakovsky CM, Logan JA, Montzka SA, Balkanski YJ, Foreman-Fowler M, Jones DBA, Horowitz LW, Fusco AC, Brenninkmeijer CAM, Prather MJ, Wofsy SC, McElroy MB (2000) Three-dimensional climatological distribution of tropospheric OH: Update and evaluation. J Geoph Res 105(D7):8931–8980

Stier P, Feichter J, Kinne S, Kloster S, Vignati E, Wilson J, Ganzeveld L, Tegen I, Werner M, Balkanski Y, Schulz M, Boucher O, Minikin A, Petzold A (2005) The aerosol-climate model ECHAM5-HAM. Atmos Chem Phys 5:1125–1156

Tanabe S, Tatsukawa R (1983) Vertical transport and residence time of chlorinated hydrocarbons in the open ocean water column. Journal of the Oceanographical Society of Japan 39:53–62

Tiedtke M (1989) A comprehensive mass flux scheme for cumulus parameterization in large scale models. Mon Wea Rev 117:1779–1800

Tomasic V, Chittofrati A, Kallay N (1995) Thermodynamic properties of aqueous solutions of perfluorinated ionic surfactants. Colloids and Surfaces, Physicochemical and Engeneering Aspects 104:95–99

Tompkins A (2002) A prognostic parameterization for the subgrid-scale variability of water vapor and clouds in large-scale models and its use to diagnose cloud cover. J Atmos Sci 59:1917–1942

Turusov V, Rakitsky V, Tomatis L (2002) Dichlorodiphenyltrichloroethane (DDT): ubiquity, persistence, and risks. Environmental Health Perspectives 101:125–128

UNEP (2001) Stockholm convention on persistent organic pollutants. http://chmpopsint/

US EPA (2005) Draft Risk Assessment of the Potential Human Health Effects Associated With Exposure to Perfluorooctanoic Acid and Its Salts. EPA Science Advisory Board (SAB)

Valcke S, Caubel A, Vogelsang R, Declat D (2004) OASIS3 User's Guide. PRISM Report Series No2:60pp

Villa S, Vighi M, Maggi V, Finizio A, Bolzacchini E (2003) Historical trends of organochlorine pesticides in an Alpine Glacier . Atmos Chem 46:295–311

Voldner EC, Li YF (1995) Global usage of selected persistent organochlorines. Sci Tot Environ 160/161:201–210

Wakeham SG, Farrington JW, Gagosian RB, Lee C, DeBaar H, Nigrelli GE, Tripp BW, Smith SO, Frew NM (1980) Organic-matter fluxes from sediment traps in the equatorial Atlantic-Ocean. Nature 286:798–800

Wallington TJ, Hurley MD, Xia J, Wuebbles DJ, Sillman S, Ito A, Penner JE, Ellis DA, Martin J, Mabury SA, Nielsen OJ, Andersen MPS (2006) Formation of $C_7F_{15}COOH$ (PFOA) and other perfluorocarboxylic acids during the atmospheric oxidation of 8:2 Fluorotelomer Alcohol. Environ Sci Technol 40:924–930

Wania F, Mackay D (1993) Global fractionation and cold condensation of low volatility organochlorine compounds in polar regions. Ambio 22:10–18

Wania F, Mackay D (1995) A global distribution model for perstitent organic chemicals. The Science of the Total Environment 160/161:211–232

Wania F, Mackay D (1996) Tracking the distribution of persistent organic pollutants. Environ Sci Technol 30(9):A390–A396

Whitman WG (1923) The two-film theory of gas absorption. Chem Metall Eng 29:146–148

WHO (2006) Indoor residual spraying - Use of indoor residual spraying for scaling up global malaria control and elimination, World Health Organization Position Paper

Wiberg PL, Harris CK (2002) Desorption of p,p'-DDE from sediment during resuspension events on the Palos Verdes shelf, California: a modeling approach. Continental Shelf Research 22:1005–1023

Wolff JE, Maier-Reimer E, Legutke S (1997) The Hamburg Ocean Primitive Equation Model HOPE. Tech. rep., German Climate Computer Center (DKRZ), Hamburg

Wolff R, Yaeger L (1993) Visualization of natural phenomena. Springer, New York, USA

Yamashita N, Taniyasu S, Petrick G, Wei S, Gamo T, Lam PKS, Kannan K (2008) Perfluorinated acids as novel chemical tracers of global circulation of ocean waters. Chemosphere 70:1247–1255

Index

About the International Max Planck Research School for Maritime Affairs at the University of Hamburg

The International Max Planck Research School for Maritime Affairs at the University of Hamburg was established by the Max Planck Society for the Advancement of Science, in co-operation with the Max Planck Institute for Foreign Private Law and Private International Law (Hamburg), the Max Planck Institute for Comparative Foreign Public Law and International Law (Heidelberg), the Max Planck Institute for Meteorology (Hamburg) and the University of Hamburg. The School's research is focused on the legal, economic, and geophysical aspects of the use, protection, and organization of the oceans. Its researchers work in the fields of law, economics, and natural sciences. The School provides extensive research capacities as well as its own teaching curriculum. Currently, the School has 15 Directors who determine the general work of the School, act as supervisors for dissertations, elect applicants for the School's PhD-grants, and are the editors of this book series:

Prof. Dr. Dr. h.c. Jürgen Basedow is Director of the Max Planck Institute for Foreign Private Law and Private International Law; *Prof. Dr. Peter Ehlers* is the Director of the German Federal Maritime and Hydrographic Agency; *Prof. Dr. Dr. h.c. Hartmut Graßl* is Director emeritus of the Max Planck Institute for Meteorology; *Prof. Dr. Lars Kaleschke* is Junior Professor at the Institute of Oceanography of the University of Hamburg; *Prof. Dr. Hans-Joachim Koch* is Managing Director of the Seminar of Environmental Law at the University of Hamburg; *Prof. Dr. Rainer Lagoni* is Director emeritus of the Institute of Maritime Law and the Law of the Sea at the University of Hamburg; *PD Dr. Gerhard Lammel* is Senior Scientist at the Max Planck Institute for Meteorology; *Prof. Dr. Ulrich Magnus* is Managing Director of the Seminar of Foreign Law and Private International Law at the University of Hamburg; *Prof. Dr. Peter Mankowski* is Director of the Seminar of Foreign and Private International Law at the University of Hamburg; *Prof. Dr. Marian Paschke* is Managing Director of the Institute of Maritime Law and the Law of the Sea at the University of Hamburg; *PD Dr. Thomas Pohlmann* is Senior Scientist at the Centre for Marine and Climate Research and Member of the Institute of Oceanography at the University of Hamburg; *Dr. Uwe Schneider* is Assistant Professor at the Research Unit Sustainability and Global Change of the University of Hamburg; *Prof. Dr. Jürgen Sündermann* is Director emeritus of the Centre for Marine and Climate Research at the University of Hamburg; *Prof. Dr. Rüdiger Wolfrum* is Director at the Max Planck Institute for Comparative Foreign

Public Law and International Law and a judge at the International Tribunal for the Law of the Sea; *Prof. Dr. Wilfried Zahel* is Professor emeritus at the Centre for Marine and Climate Research of the University of Hamburg.

At present, *Prof. Dr. Dr. h.c. Jürgen Basedow* and *Prof. Dr. Ulrich Magnus* serve as speakers of the International Max Planck Research School for Maritime Affairs at the University of Hamburg.